AF730613

Marie's Computer Notes

Microsoft® Excel™ Data Validation

by Marie Eklof

Marie's Computer Notes: Microsoft® Excel™ Data Validation

CONTENTS

FORWARD

There are several reasons why I wrote this book. I wanted to have my computer notes easily accessible so that I could use them myself. I also wish I would have known about data validation when I started to work in Excel, since it has improved the quality of my work and saves me time. I want other people to know how easy it is. I have a very large Excel spreadsheet that I work with at my job. I took this sheet over from someone else before I knew about data validation. Every time I look at the way company names are spelled in so many various ways, I am dismayed. Now I even use data validation for my personal Excel spreadsheets, recording the categories of charges on my credit card.

I work in international trade. I encourage my clients to submit paperwork that is easy to understand. I will assist them if they allow it. When I casually ask them if they use Excel, I can sense that some of them are afraid of it. They do not want to clean up their paperwork because they do not know where to begin. In this book, I will show you how to create a simple Excel spreadsheet from the very beginning and how to keep your Excel spreadsheet clean with Excel data validation. When you do business with any company and you are required to send in paperwork, you will receive better customer service if the paperwork is clear and concise.

Data validation is a quality control feature you can use to increase the accuracy of your data entry. You can apply data validation to either numbers or text. This ensures that your numerical data is logical. For text, you can use data validation to create a drop-down from a list you create, ensuring that your data is spelled correctly.

I wrote this book not knowing how much experience the reader has in creating a spreadsheet. I wanted to show the reader how to create an

Excel spreadsheet, before showing them how to use data validation in their spreadsheets.

When I speak with my clients and casually ask them if they work in Excel, I can sense that some of these people are afraid of Excel. I want to show people that creating an Excel spreadsheet, is not difficult, if you are shown how to do it.

In this book you will learn how to:

- Create an Excel spreadsheet.
- Set up data validation for quality control.
- Filter and sort spreadsheet data.
- Add to the data validation list.
- Change the storage column information.
- Remove data validation.
- Apply what you have learned.

CHAPTER 1: SETTING UP YOUR SPREADSHEET

In this chapter, we discuss how to set up a spreadsheet, including a discussion of the following Excel features:

- Creating an Excel spreadsheet
- Changing the color of the spreadsheet
- Changing the borders on a spreadsheet
- Changing the size of the font on a spreadsheet
- Wrapping text so that it fits into the cell of a spreadsheet
- Freezing panes so you can navigate through your spreadsheet more easily
- Saving an Excel file

CREATING THE MICROSOFT EXCEL SPREADSHEET

Let's start by creating the Excel spreadsheet.

1. To start a spreadsheet, open Excel. You can either click the Excel icon on the desktop or search for Excel using the search box in the lower-left corner of your screen.

   Excel opens with options of how to create the spreadsheet.

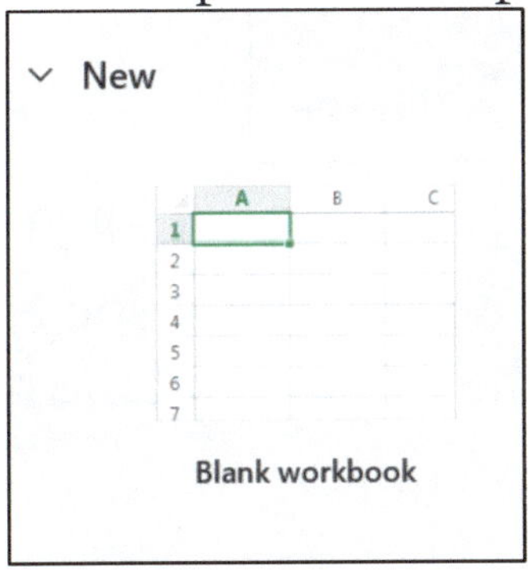

2. Select the option to create a blank workbook.

3. Click anywhere in the spreadsheet.
4. Now create your header row. For this spreadsheet, we will use Cells A1 through D1, highlighted below, which will become your header row.

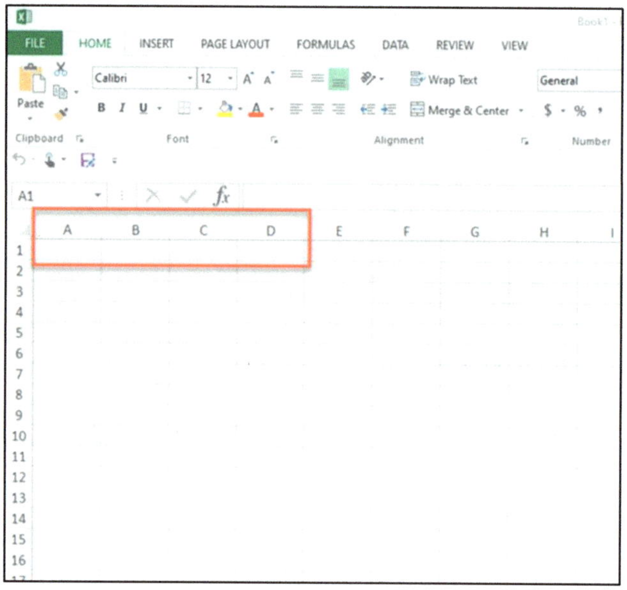

5. Enter the following text for each of the cells in the header row.
Cell A1: **Food Type**
Cell B1: **Beverage Type**
Cell C1: **Alcohol Type**
Cell D1: **Species of Mushrooms**
6. Widen all the columns above so that the text does not run over to an adjacent cell. For this example, let's make Cell D1 (with Species of Mushroom text) wider. To do this, place your cursor on the line that separates the Cells D1 and E1.
The cursor turns into a "Column/Row Resize" mouse pointer.

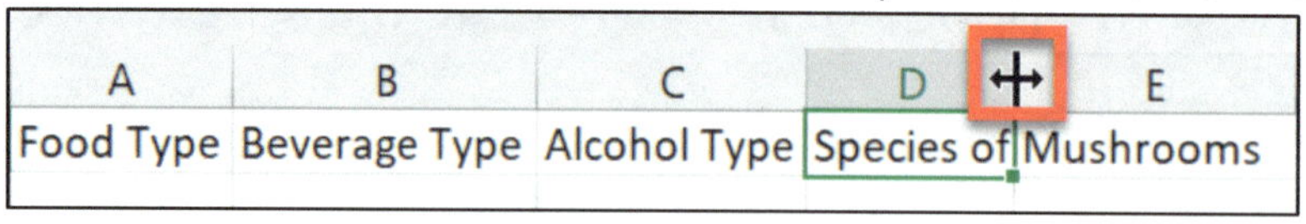

7. Place your cursor on the mouse pointer and widen Cell D1 so "Species of Mushrooms" will fit into one cell. Once you have clicked onto the mouse pointer, drag the cursor until the row is wide enough.

Tip: *A shortcut to widen the cells is to double-click on the Column/Row Resize mouse pointer itself.*

Now you can see that "Species of Mushrooms" fits into Cell D1. That is, the text does not "spill over" to Cell E1 anymore.

	A	B	C	D
1	Food Type	Beverage Type	Alcohol Type	Species of Mushrooms
2				

SAVING THE SPREADSHEET

When you have finished with a task it is important to save. The first time you save your file, you should also name it.

To save your file, click **File** > **Save**. A shortcut to save the file is to simply click inside the spreadsheet and press **Ctrl-S**.

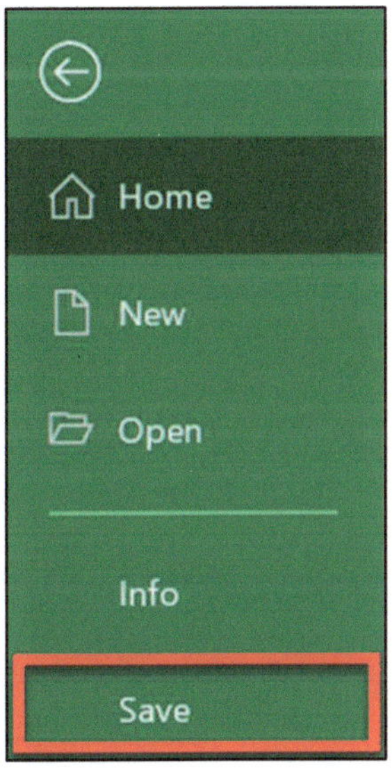

Since this is the first time you have saved your spreadsheet, the Save As dialog box appears, prompting you to name the file. Let's name the file *Products.xlsx*.

Next, we will go over several ways to cosmetically improve your spreadsheet including changing the color, font, font size, header row, and borders.

CHANGING THE COLOR OF THE SPREADSHEET

Now we will change the color of the spreadsheet.

1. Place your cursor on the left corner of the spreadsheet, where I have the red square below.
   This selects and highlights the whole spreadsheet.

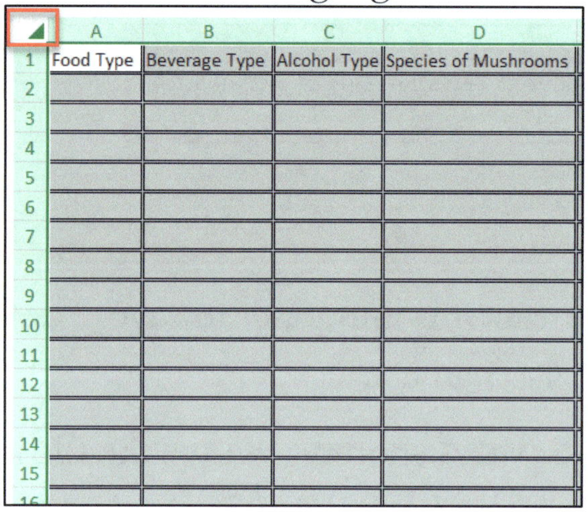

2. On the top menu bar, select the **Home** menu.
3. Click the arrow next to **Fill Color** ().

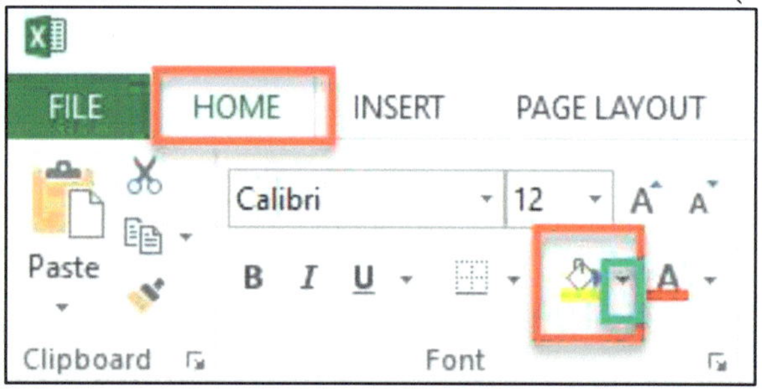

The color options appear.

4. Select **More Colors**.

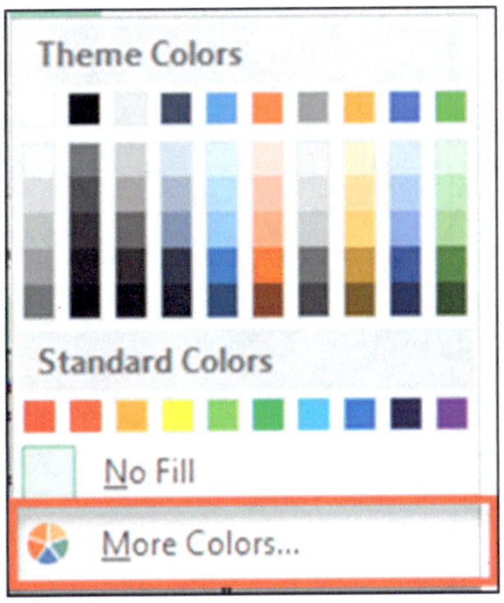

5. Select a color by clicking on it.

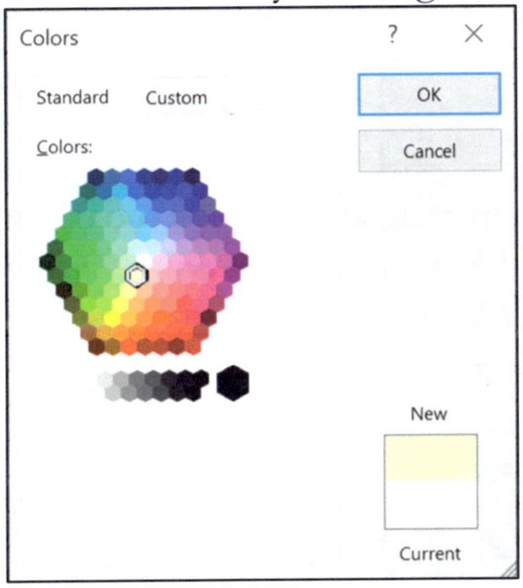

Now we will customize the shade of this color.

6. Click the **Custom** tab.

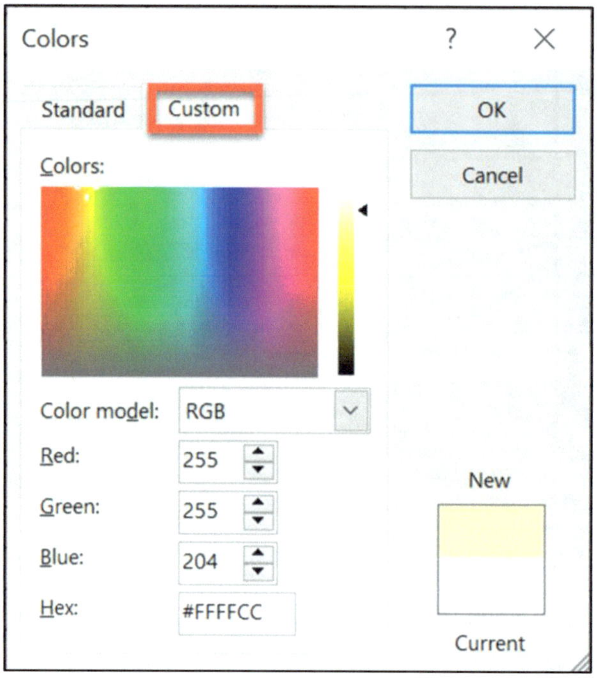

7. Make the color lighter by moving the arrow up. To darken the color, move the arrow down. I have made the color darker.

8. Click **OK**.

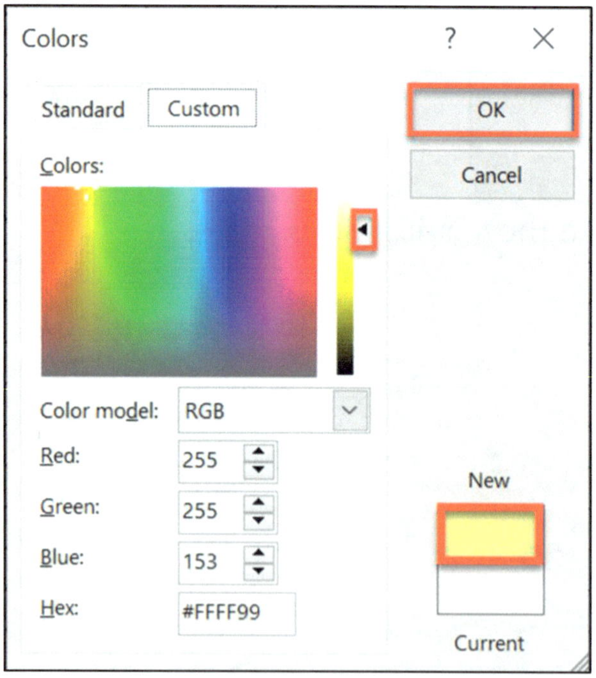

Now you can see that the color of your spreadsheet has changed.

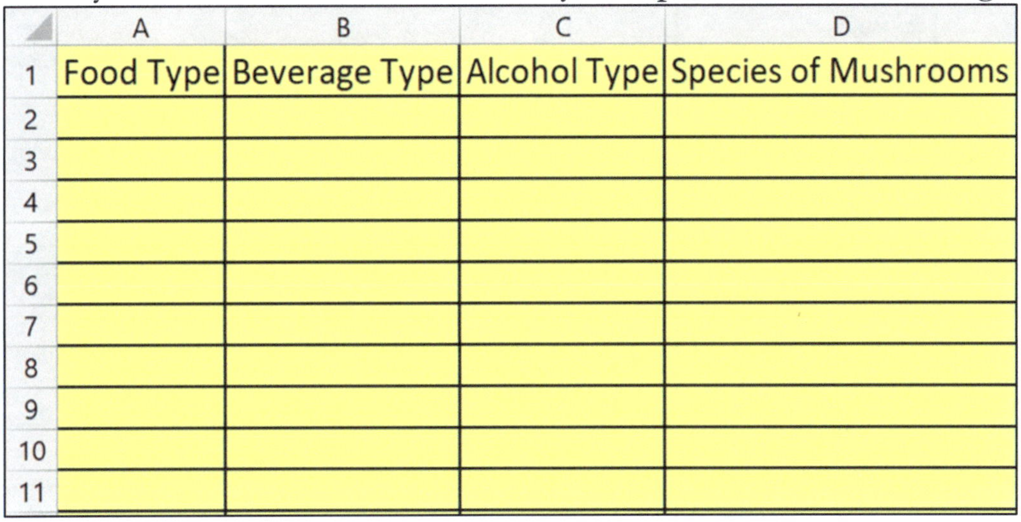

CHANGING THE BORDER LINE STYLE OF THE SPREADSHEET

Now I will show you how to change the borders on the spreadsheet to a new line style.

1. Place your cursor in the top-left corner cell of the spreadsheet. This selects the whole spreadsheet.

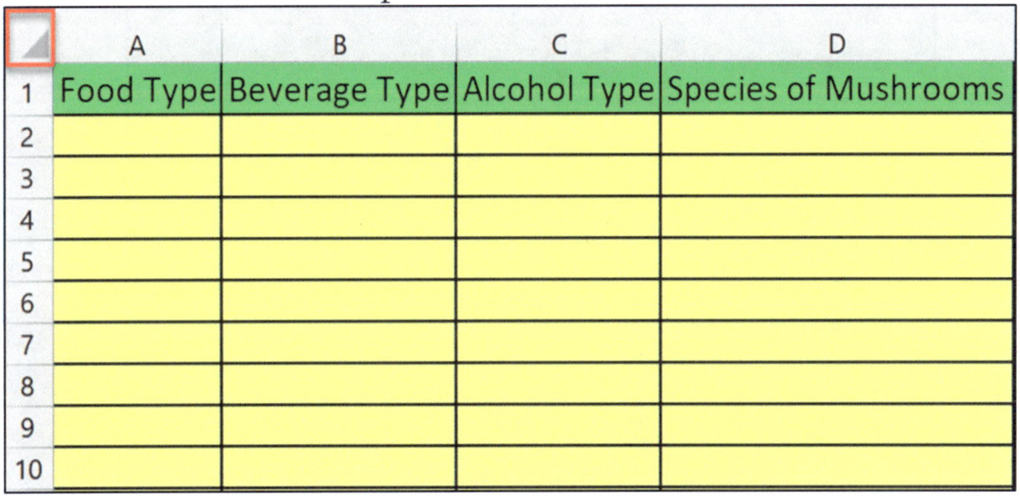

2. On the top menu bar, select the **Home** menu and then **Format**.

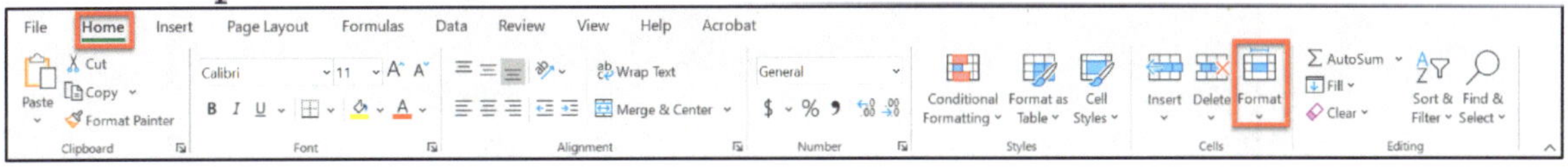

The Format options appear.

3. Select **Format Cells.**

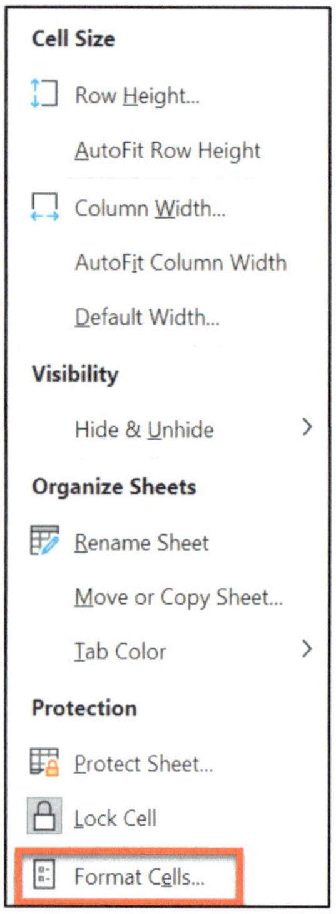

The Format Cells dialog box appears.

4. Select **Border.**

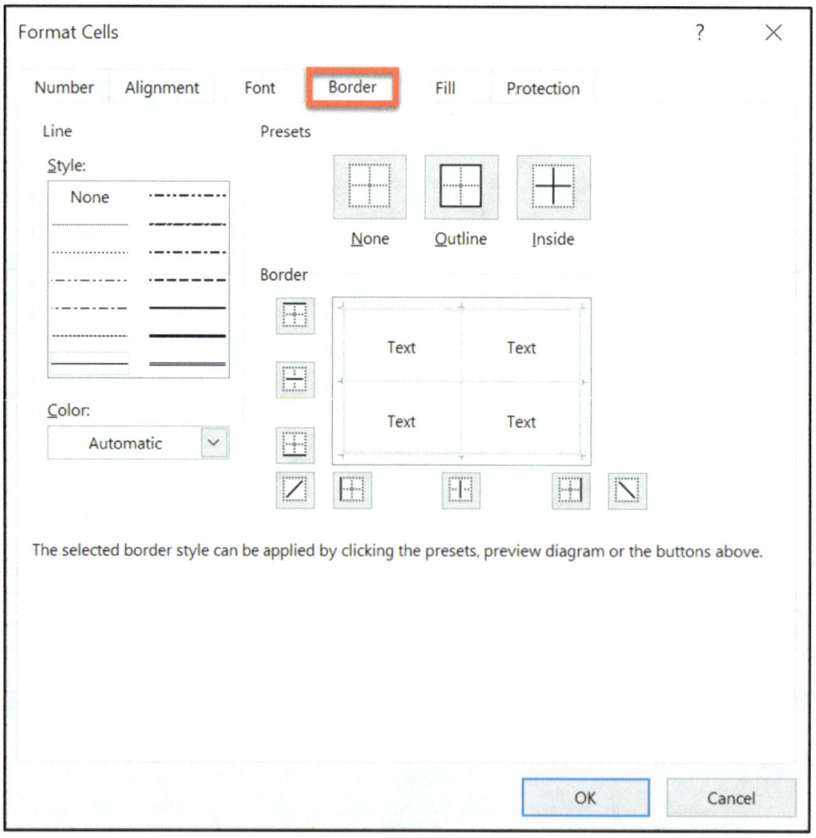

5. Click the line style that you want to appear in the spreadsheet. Let's select the double lines for this exercise.

6. Click on the **Outline** and **Inside** boxes.

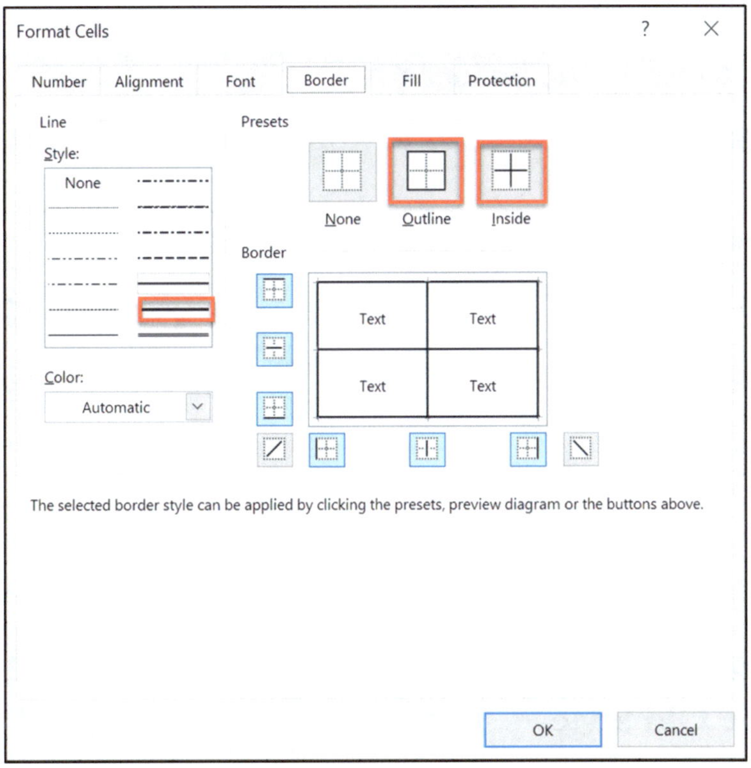

You can see that the borders changed.

7. Click **OK**.

Now you can see that the borders have changed on the spreadsheet.

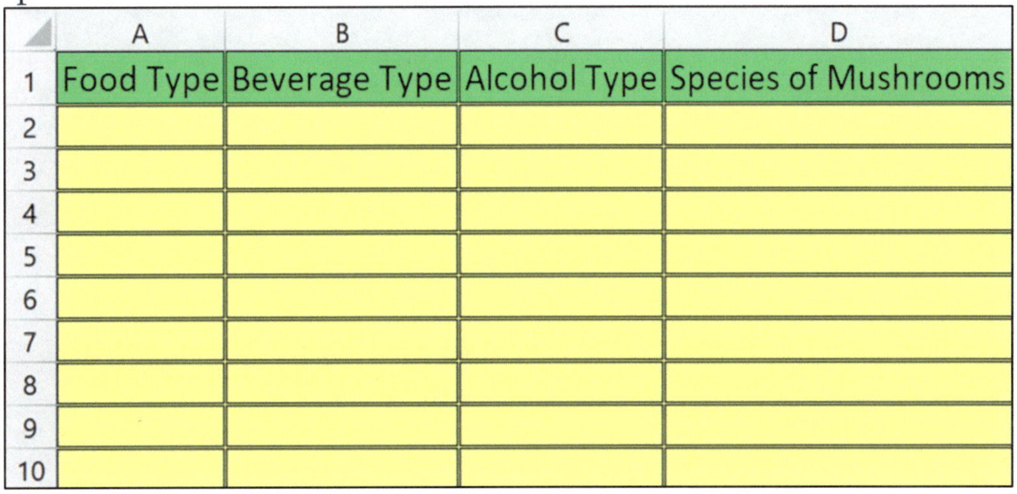

Changing the Font Style of the Header Row

To show you how to change the font, we will temporarily change the font style of the header row, which is Row 1.

1. Click Row 1 to select the header row.
2. Select the arrow to the right of the word Calibri. This is the font selector.

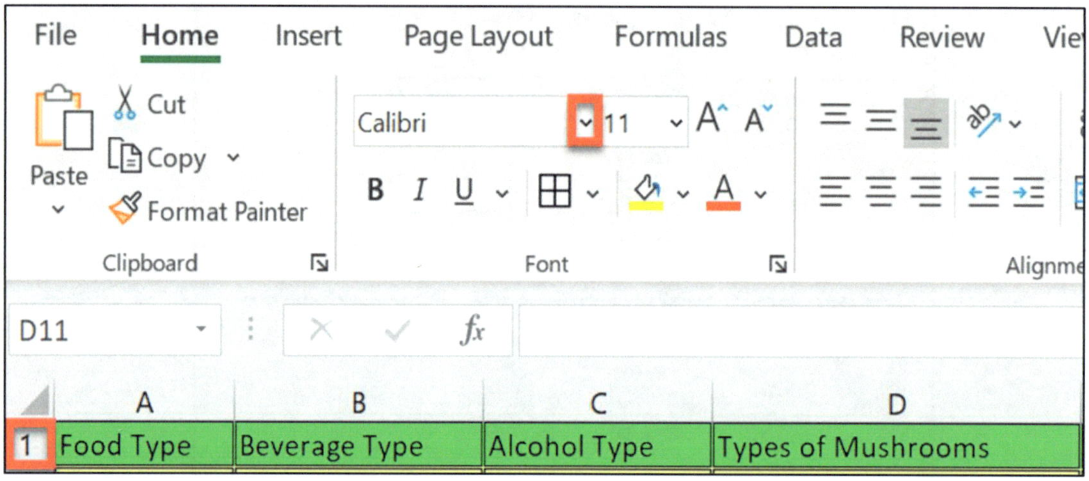

3. Select **Goudy Old Style**.

4. To move the drop-down so you can see more fonts, click the scrollbar on the side of the page, and move it up or down.

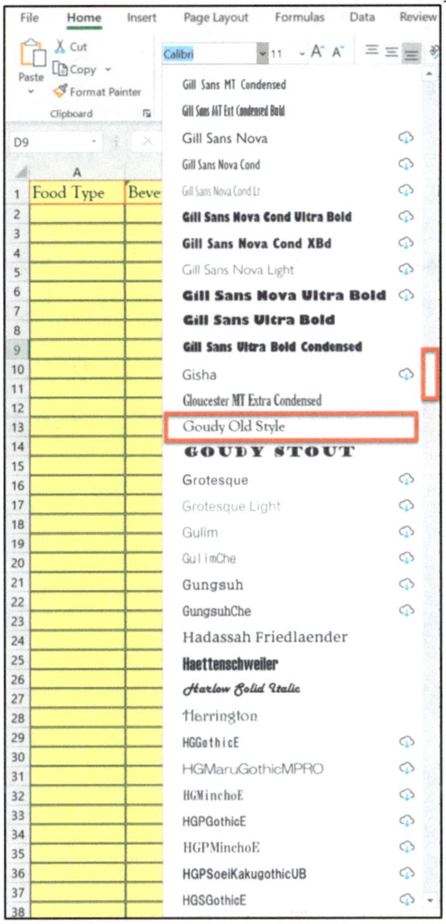

Now you can see that the font of the header row has changed.

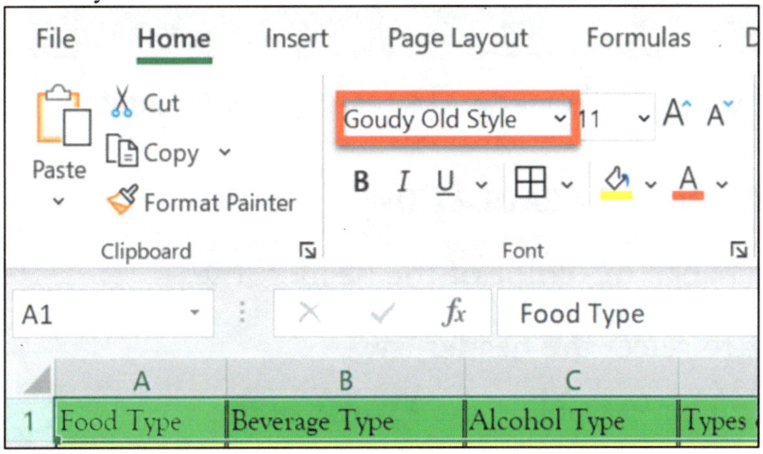

5. Click in the spreadsheet and press **Ctrl-S** to save it.

CHANGING THE FONT SIZE OF THE HEADER ROW

Now we will change the font size of the header row.

1. Click Row 1 to select the header row.
2. On the top menu bar, select the **Home** menu if it is not already selected.
3. Change the font from Old Goudy Style back to Calibri so that you can follow along with the procedures in the rest of this book.
4. Click the drop-down arrow next to 11. This is the font size selector. Select **14** from the list. You can also simply type in a new size on top of the old size.

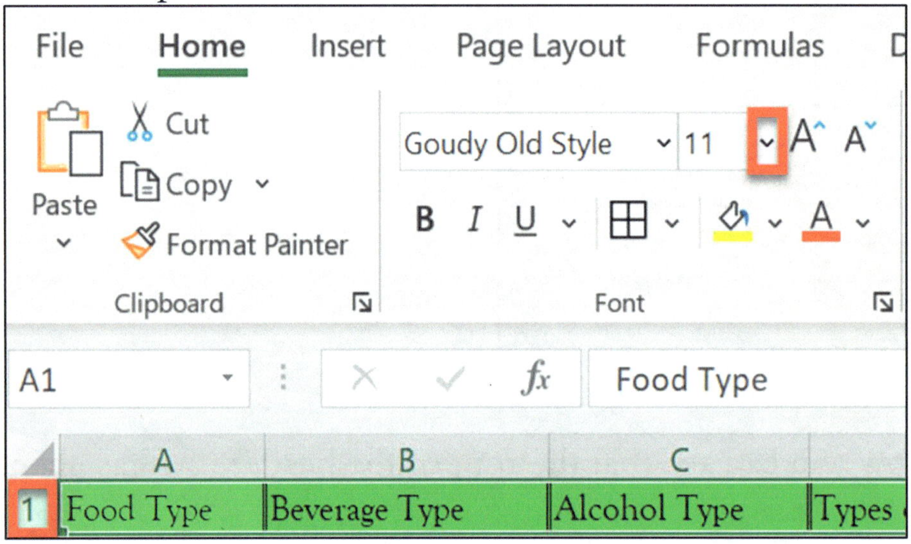

5. Click inside the spreadsheet and press **Ctrl-S** to save.

CHANGING THE COLOR OF THE HEADER ROW

Now we want to change the color of the header row.

1. Click Row 1 to select the header row.
2. On the top menu bar, select the **Home** menu.
3. Click the drop-down arrow next to . This is the fill color selector.

4. Click **More Colors**.

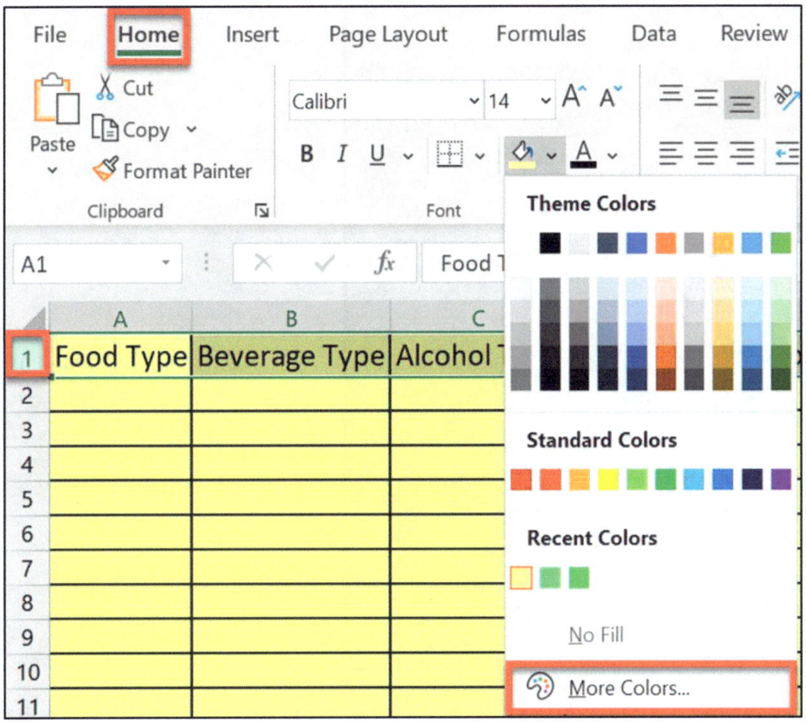

5. Click a color to select it.

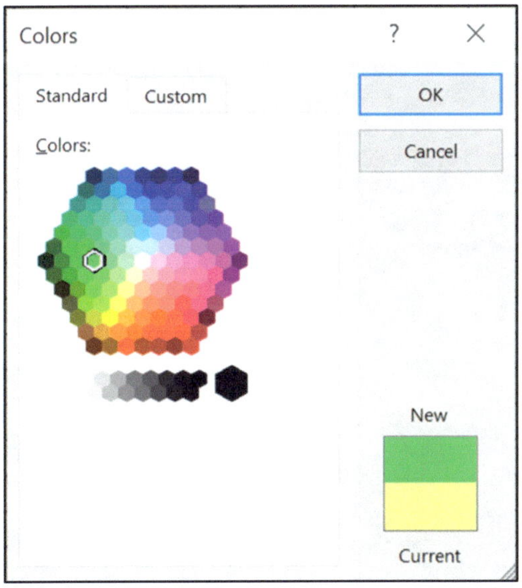

The new color shows up on the bottom right.

6. Now click the **Custom** tab.

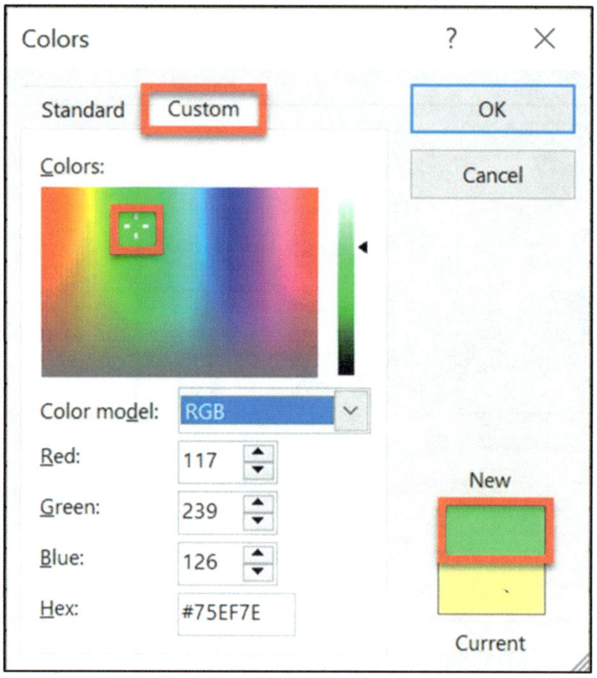

7. Make the color lighter by moving the arrow up. To darken the color, move the arrow down.
8. Once you have made your choice, click **OK**.

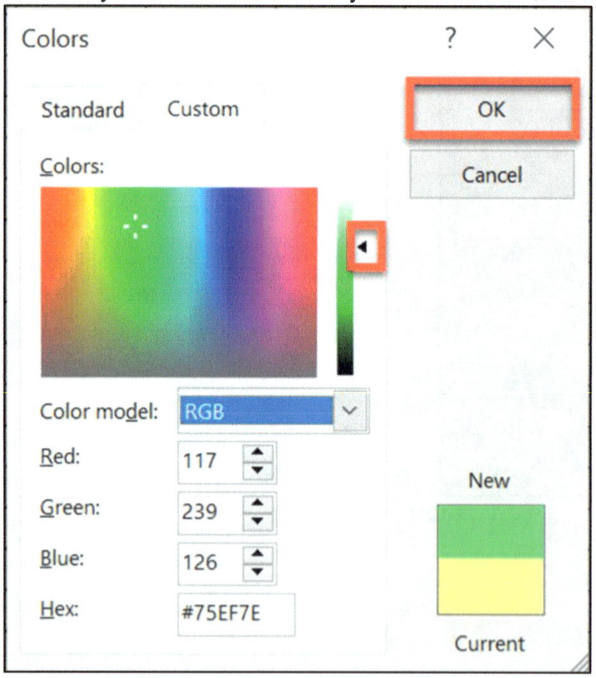

9. Now you can see that the new header row color has been applied.

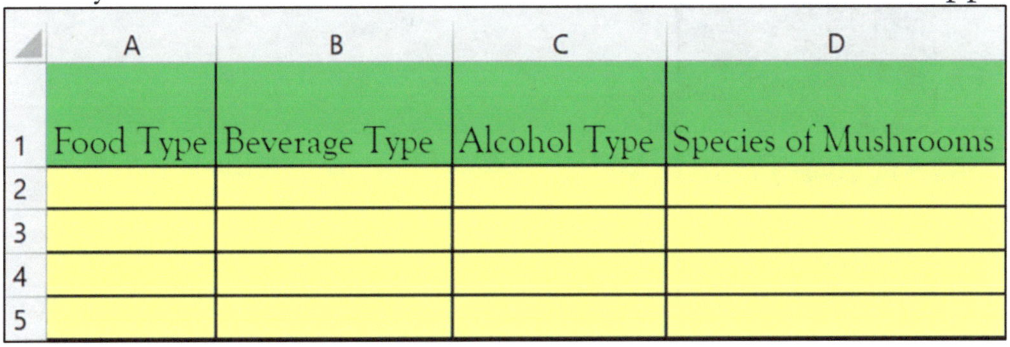

10. Click anywhere in the spreadsheet and press **Ctrl-S** to save it.

CHANGING THE BORDER LINE STYLE OF THE HEADER ROW

Now we will change the borders of the header row to make them stand out more.

1. Click Row 1 to select the header row.

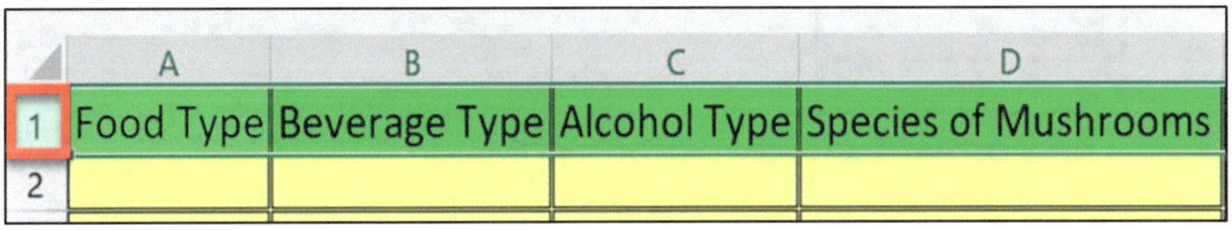

2. On the top menu bar, select the **Home** menu.

3. Select **Format** and then **Format Cells**.

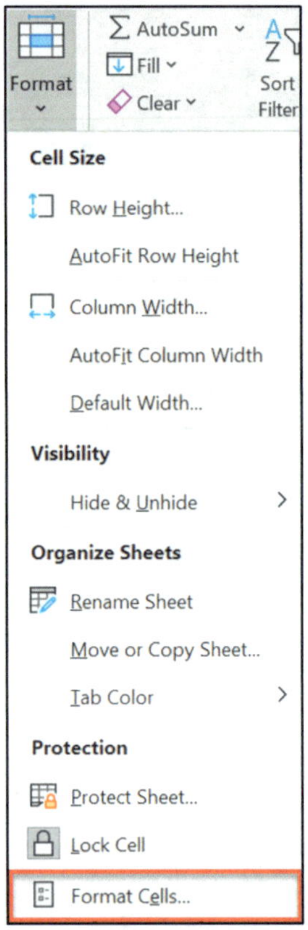

The Format Cells dialog box appears.

4. Select the **Border** tab.

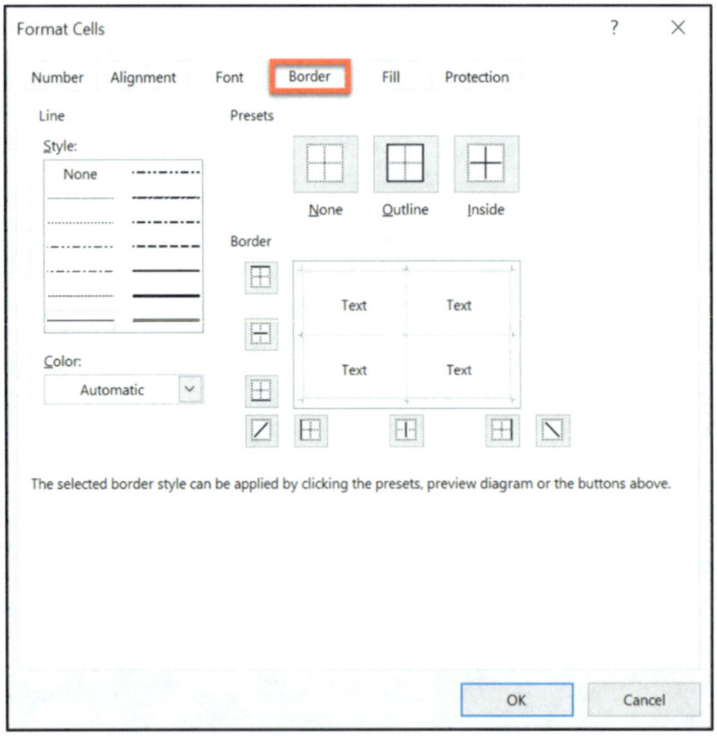

5. Select the thick black line so that the header row is darker and more prominent.

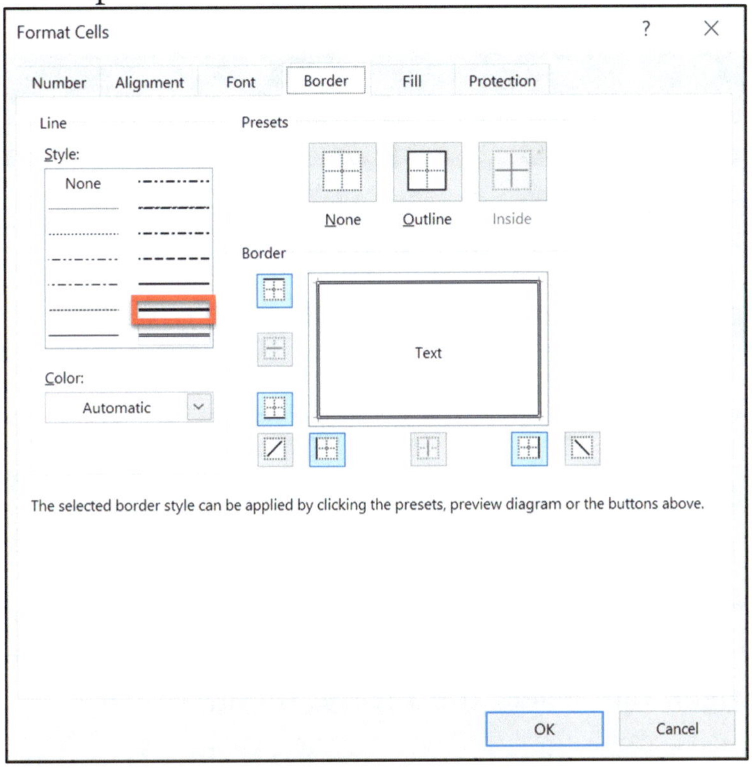

6. Click onto the **Outline** and **Inside** boxes.

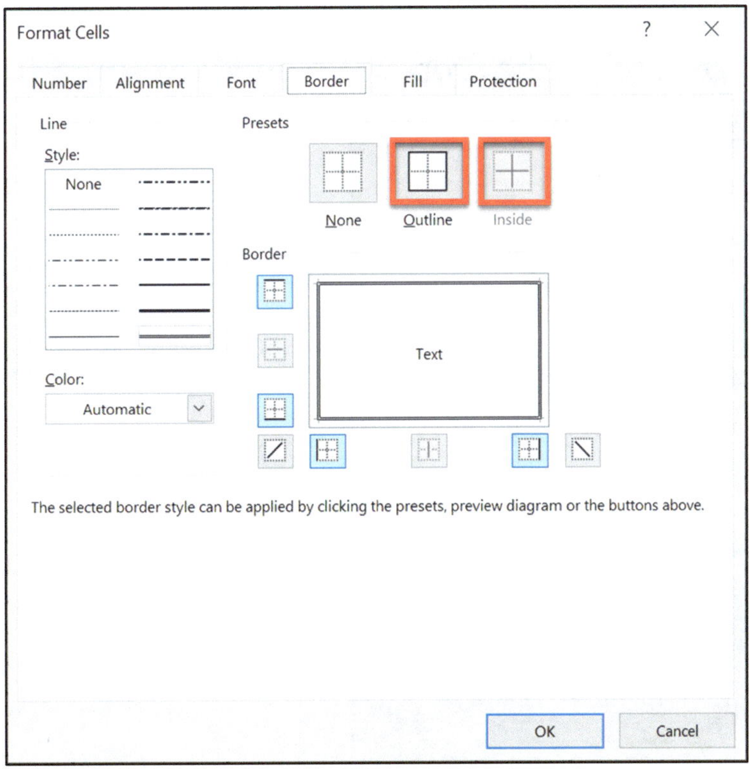

7. Click **OK**.
   You can see that the lines are now darker.

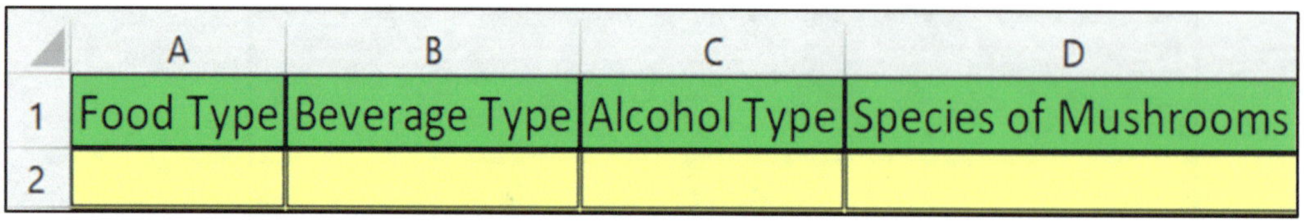

CHANGING THE BORDER COLOR OF THE HEADER ROW

Now we will change the border color on the header row.

1. Click Row 1 to select the header row.

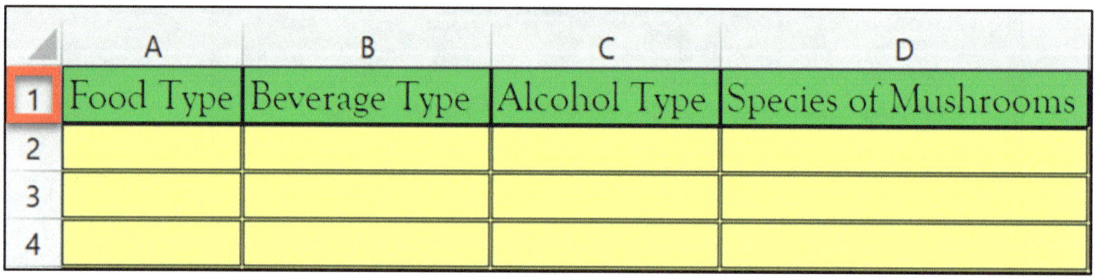

2. On the top menu bar, select the **Home** menu.
3. Select **Format**, then in the drop-down list, select **Format Cells**.
4. Select the **Border** tab.
5. In the Color area, select the drop-down under **Automatic**.
   The color picker appears.

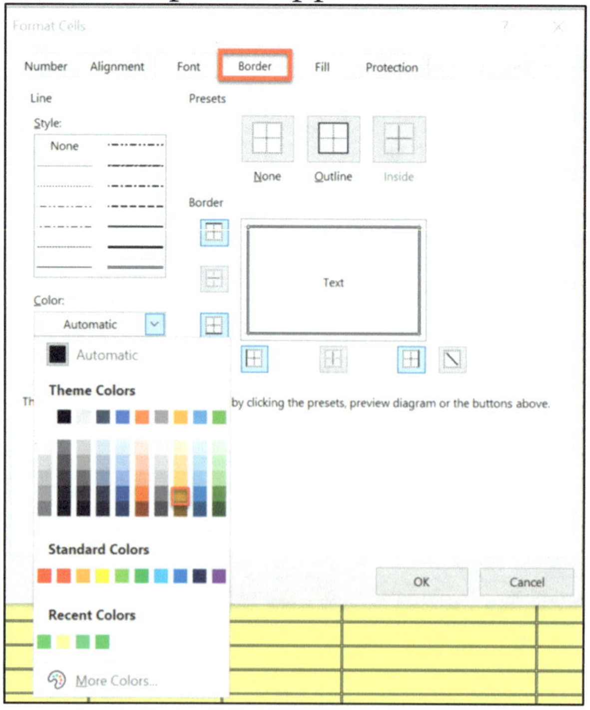

6. Select the color that you want by clicking it. For this example, let's choose brown as shown in the figure below.

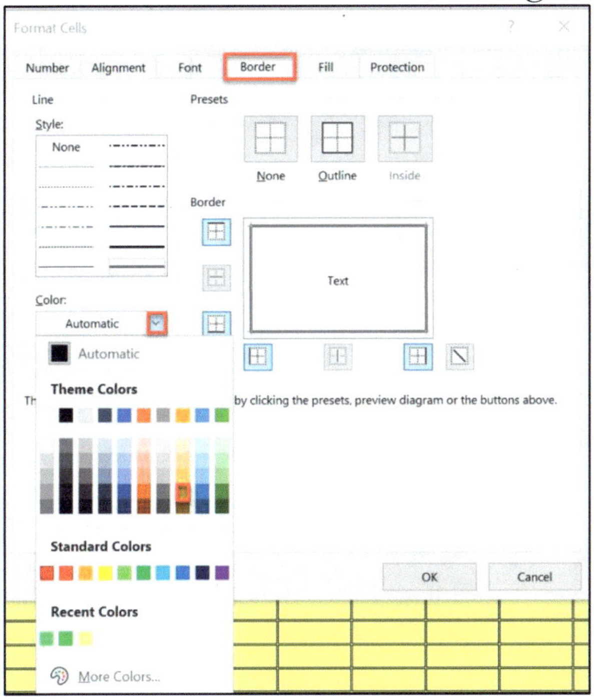

7. Click **Outline and Inside**.
   This changes the color for both the inside and outside borders of your header row.

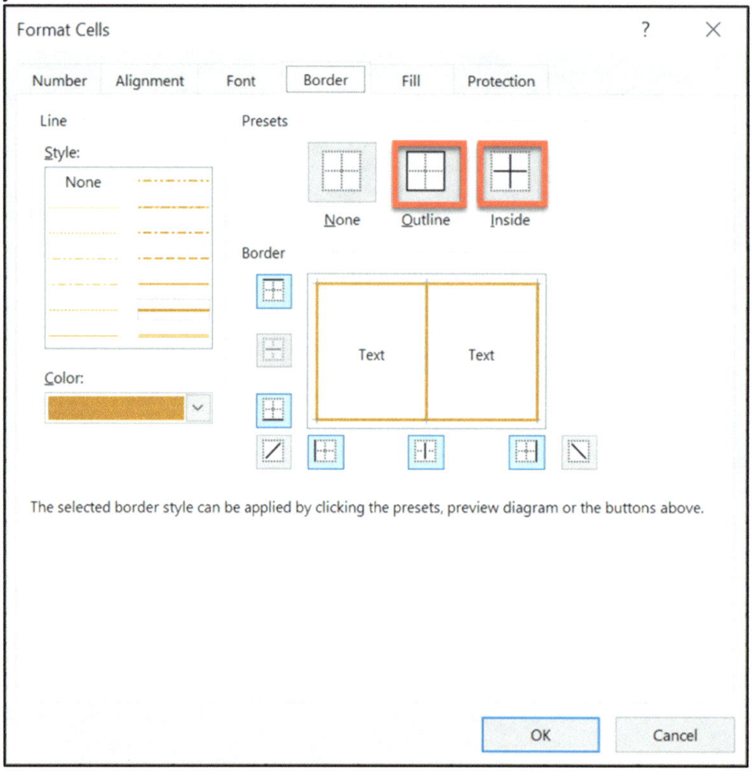

8. Click **OK**.
   Now you can see that the border color in the header row has changed color.

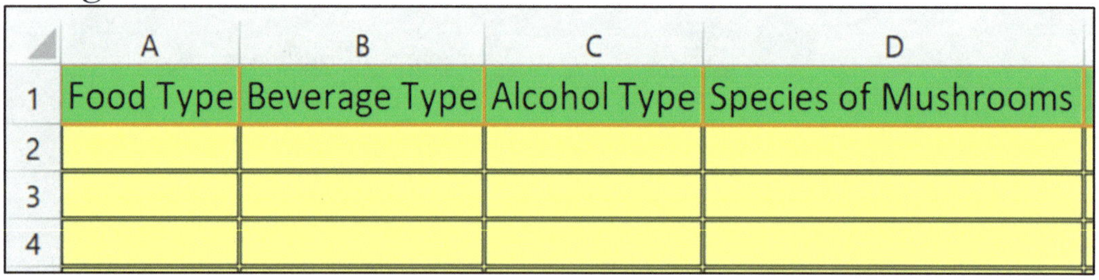

9. Click in the spreadsheet and press **Ctrl-S** to save it.

Wrapping Text

When you wrap text, it displays the cell contents on multiple lines, rather than one long line.

1. Click Row 1 to select the header row.
2. On the top menu bar, select the **Home** menu, then select **Wrap Text**.

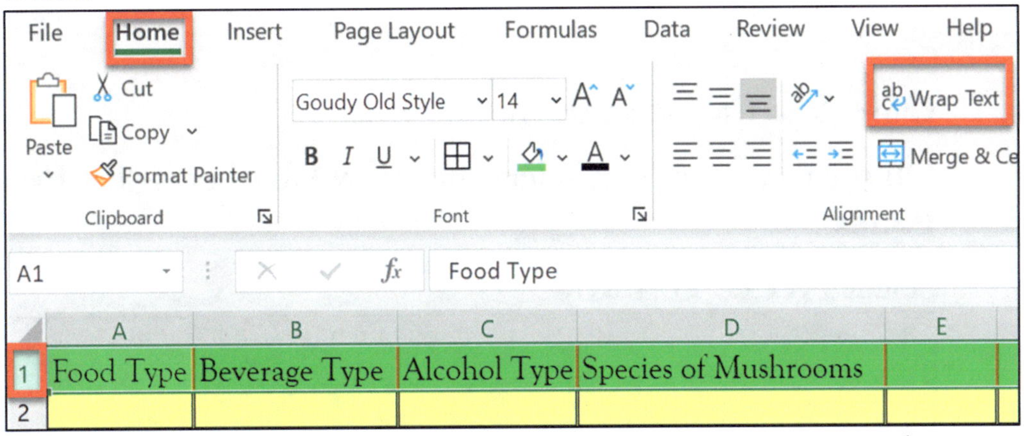

Now you can see that the heading row text wraps onto the next line.

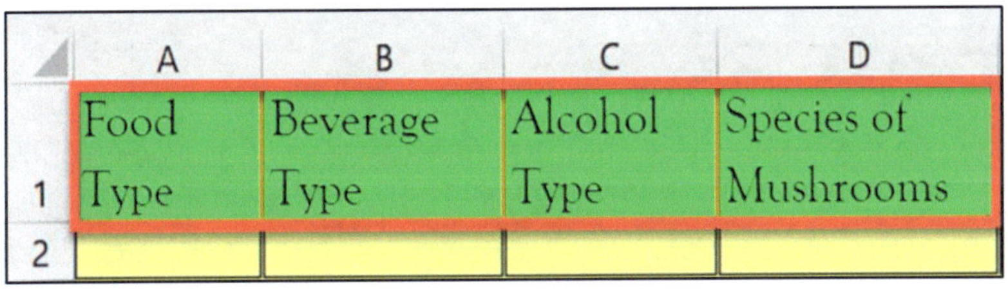

Freezing Panes

Now I will show you how to freeze panes. This feature allows you to always see the top row and column headers as you scroll through your spreadsheet.

For our example, we are going to freeze the pane so that when you move further down on the spreadsheet your header row will always be visible.

1. Click Row 1 to select the header row.
2. On the top menu bar, select the **View** tab, then select **Freeze Panes**.
3. Select **Freeze Top Row**.

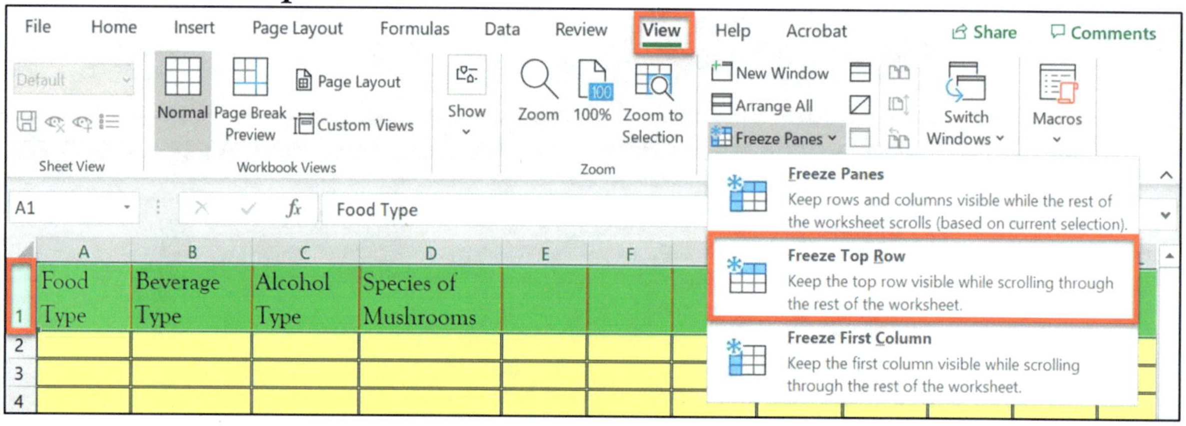

4. Click anywhere in the spreadsheet and press **Ctrl-S** to save it.

Chapter 2: Using Data Validation for Quality Control

In this chapter, we discuss the following data validation topics.

- Examples of how to use data validation in your personal life
- How to create an acceptable range of *numbers*
- How to create a list of possible options for *text*, allowing you to avoid misspellings in your spreadsheet
- How to create data validation warning and error alert messages that appear when you type in an incorrect entry

Using Data Validation in Your Personal Life

Below are a few examples of how I use data validation in my personal life.

Every week I pay off my credit card to make it easier to track my charges and avoid interest charges. I record my charges, on one spreadsheet, my payments on another, and I have the data validation in the middle sheet. Here is an example of the sheet showing my payments. It has tabs on the bottom for Payment Date, Data Validation, and Charges on Visa Card.

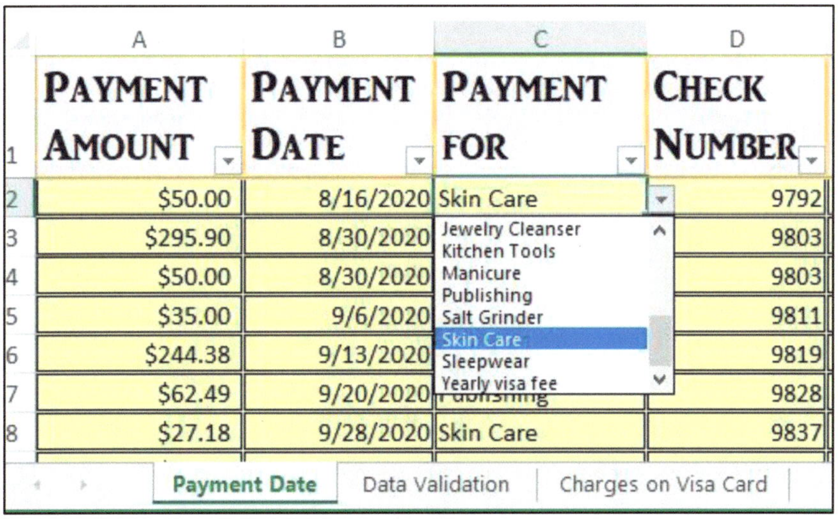

Example 1

Here is where is I record my credit card charges. By doing this, I feel that I can ensure that all the charges are accurate, have more control of my finances, and avoid paying any unnecessary interest or fees.

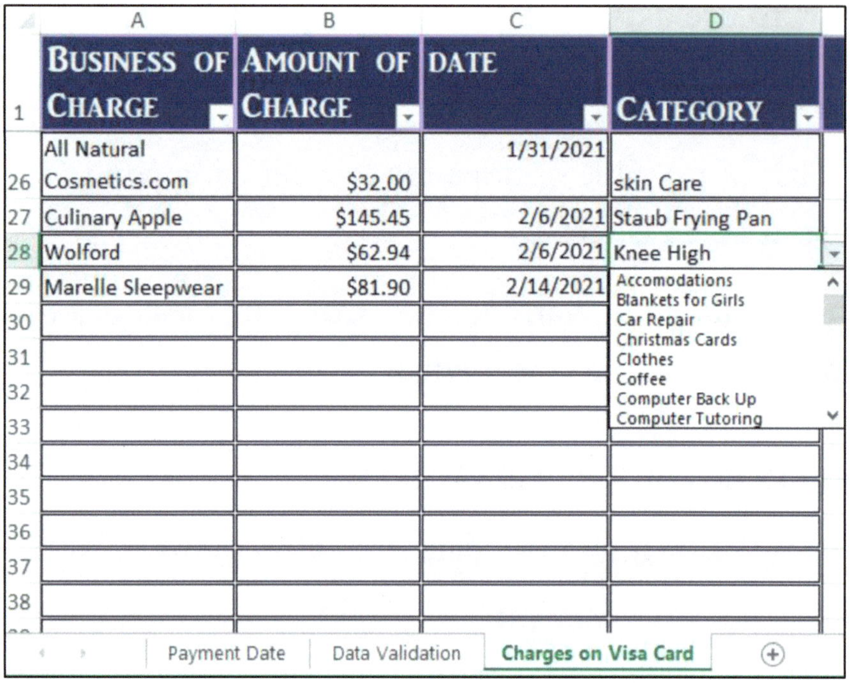

	Business of Charge	Amount of Charge	Date	Category
26	All Natural Cosmetics.com	$32.00	1/31/2021	skin Care
27	Culinary Apple	$145.45	2/6/2021	Staub Frying Pan
28	Wolford	$62.94	2/6/2021	Knee High
29	Marelle Sleepwear	$81.90	2/14/2021	
30				
31				
32				
33				
34				
35				
36				
37				
38				

EXAMPLE 2

Here is another example of how I use data validation for personal use. I have inserted data validation both into Column A and C.

Cosmetics and skin care are pricey, and I love to try different brands. If I do not record the product type and color, I will never remember them. To minimize typing and avoid errors, I use validation to create drop-down lists in Excel both for brand and product type.

Here is an example of the drop-down list for brand.

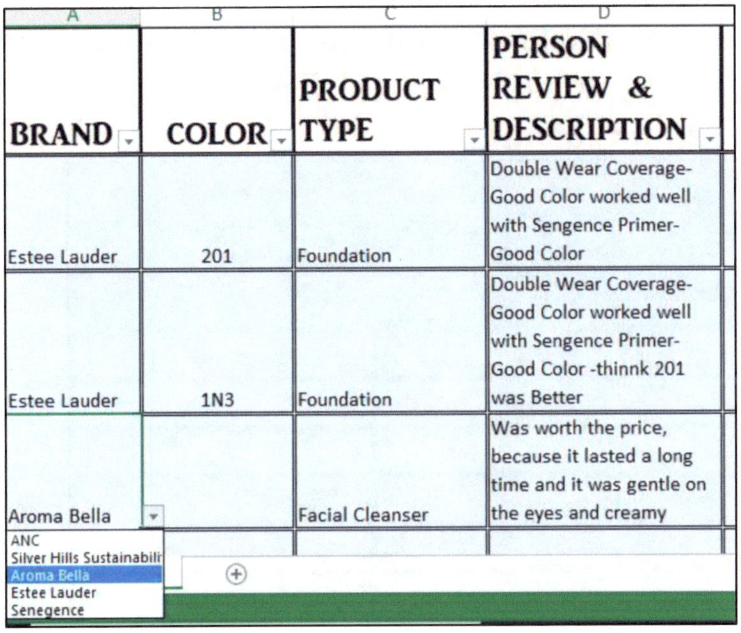

BRAND	COLOR	PRODUCT TYPE	PERSON REVIEW & DESCRIPTION
Estee Lauder	201	Foundation	Double Wear Coverage-Good Color worked well with Sengence Primer-Good Color
Estee Lauder	1N3	Foundation	Double Wear Coverage-Good Color worked well with Sengence Primer-Good Color -thinnk 201 was Better
Aroma Bella		Facial Cleanser	Was worth the price, because it lasted a long time and it was gentle on the eyes and creamy

Here is an example of the drop-down list for product type.

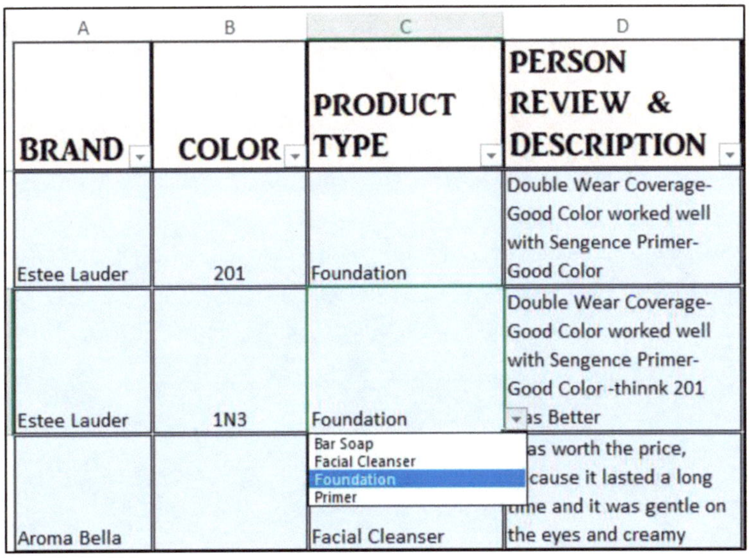

BRAND	COLOR	PRODUCT TYPE	PERSON REVIEW & DESCRIPTION
Estee Lauder	201	Foundation	Double Wear Coverage-Good Color worked well with Sengence Primer-Good Color
Estee Lauder	1N3	Foundation	Double Wear Coverage-Good Color worked well with Sengence Primer-Good Color -thinnk 201 was Better
Aroma Bella		Facial Cleanser	Was worth the price, because it lasted a long time and it was gentle on the eyes and creamy

Creating Data Validation with Numbers

The following procedure demonstrates how to create a drop-down list with valid numerical choices.

First, open the spreadsheet we created in Chapter 1.

For this scenario, let's assume that a company is inputting the Stock Keeping Unit (SKU), or product number for their beverage varieties they have in stock. All the beverage SKU product numbers begin at 1983 and stop at 3900.

1. Select the column or row where you want to insert data validation. For this example, select Column B (that is, click on "B") so that you can validate the entire column.

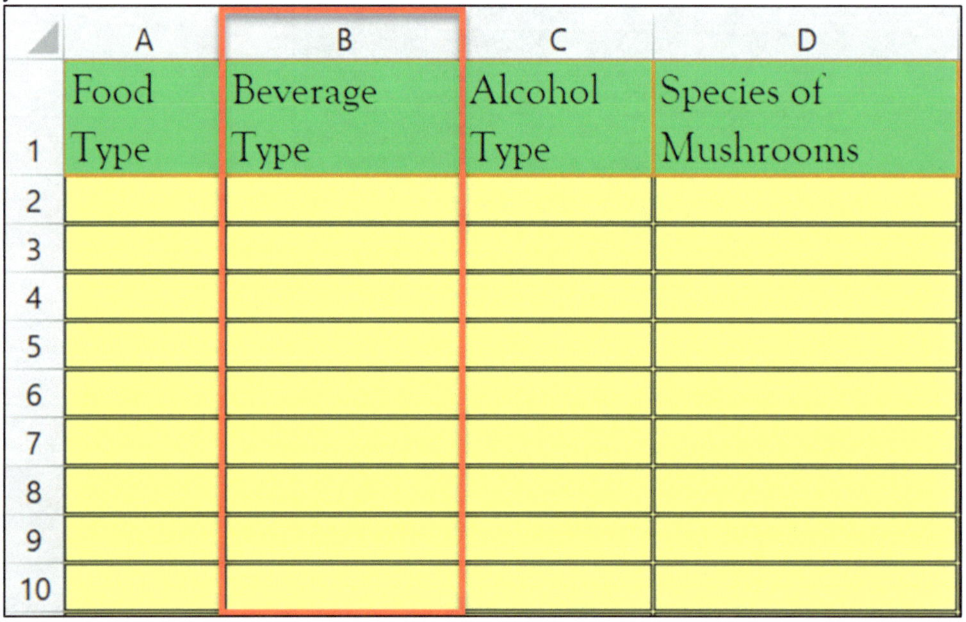

2. Select the Data menu, then 

, and then **Data Validation.**

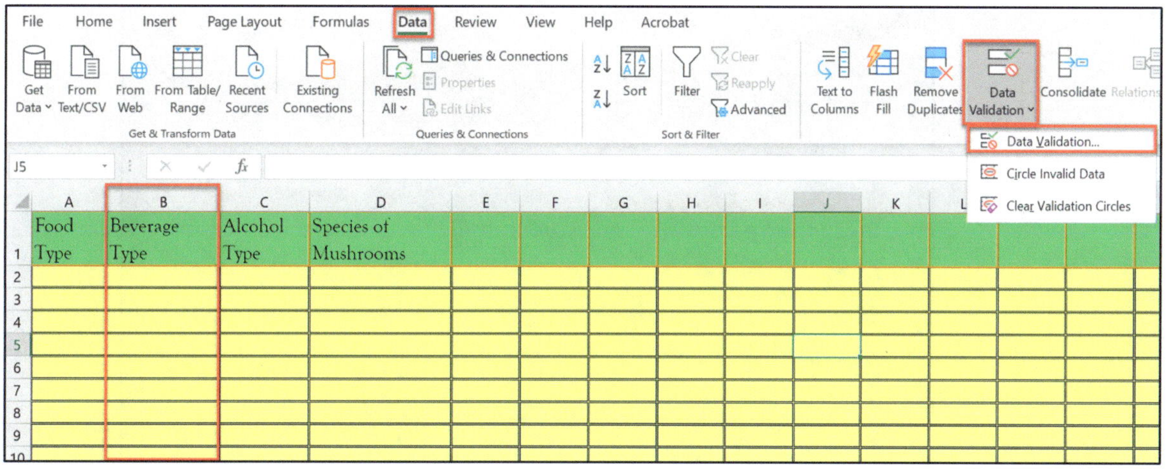

The Data Validation dialog box appears.

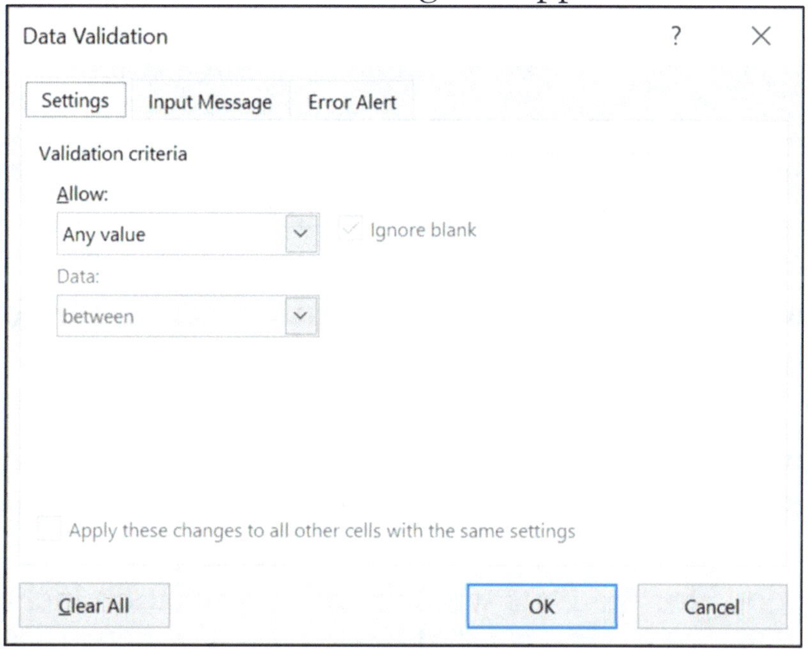

The Data Validation dialog box has three tabs: Settings, Input Message, and Error Alert.

On the Settings tab, the default option for the Validation criteria is "Any value." This means that you can type anything in Column B, and it will be allowed. You need to change the setting if you want to validate data in Column B.

3. In the Allow list, change Any Value to **Whole number** using the drop-down menu as shown below.

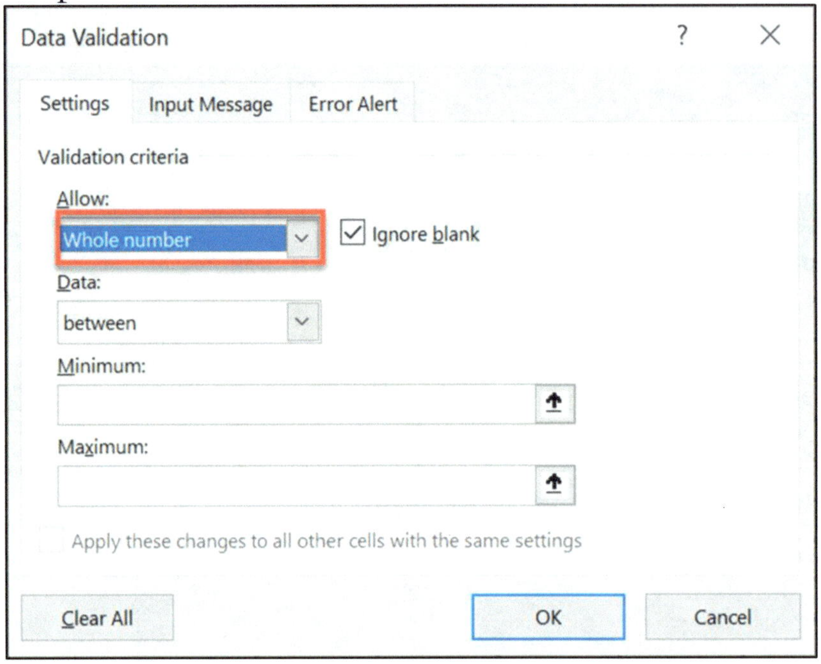

Once you change the Allow option to allow whole numbers, the Data drop-down options become available.

The default option for Data is "between." When you choose this option, you can set Minimum and Maximum numbers.

After you set these numbers, you can only enter values in the cells you selected (Column B in this example) that are between them.

Other options for the Data whole number validation include "not between," "equal to," "not equal to," "greater than," "less than," "greater than or equal to," and "less than or equal to."

4. Change the Data option to **between**.

5. Enter **1983** for the Minimum, **3900** for the Maximum, and then click **OK**.

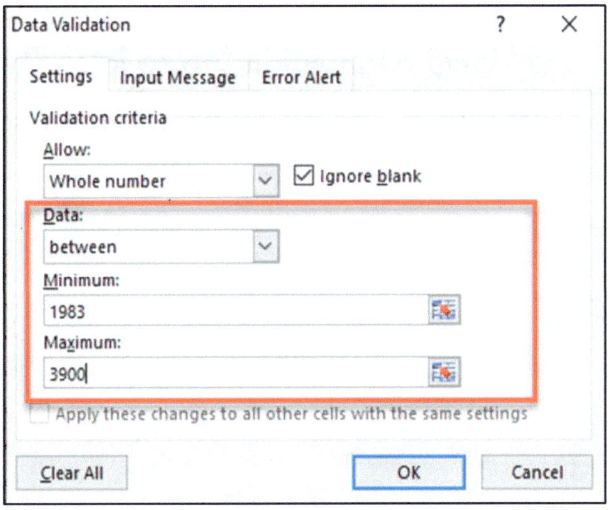

6. Now type **1982** in Cell B2.

You will find that Excel will not allow you to enter this number, since 1982 is not in the range of beverage numbers that you set up in the data validation.

You can click Retry to enter a new value or click Cancel. Let's click **Cancel**.

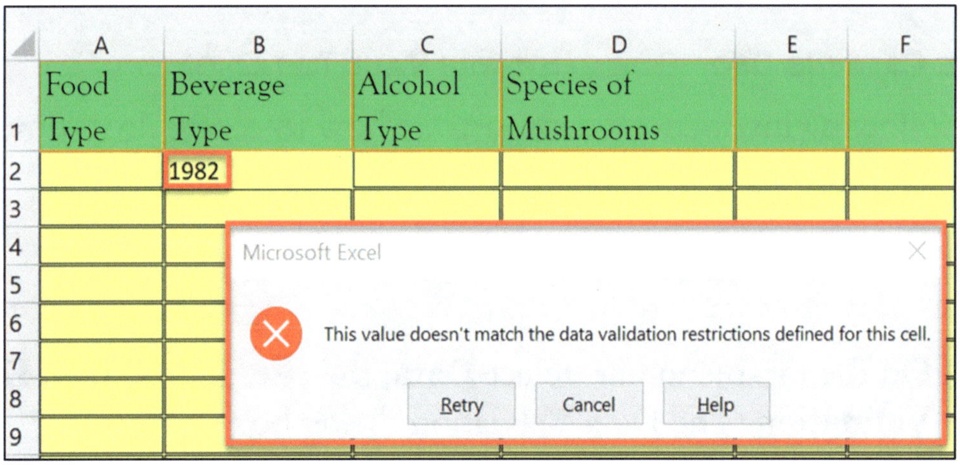

Note: Since you're applying data validation to the whole column, if you were to change a text heading such as "Beverage Type" above, you would also get an error. In other words, since the Minimum and Maximum values in this data validation are numbers, you cannot substitute text for them.

CREATING DATA VALIDATION WITH TEXT

See below how the name of one company can appear in one spreadsheet may different ways. I have used Excel data validation to keep this from happening to Excel spreadsheets where I have client names. There is even one word misspelled.

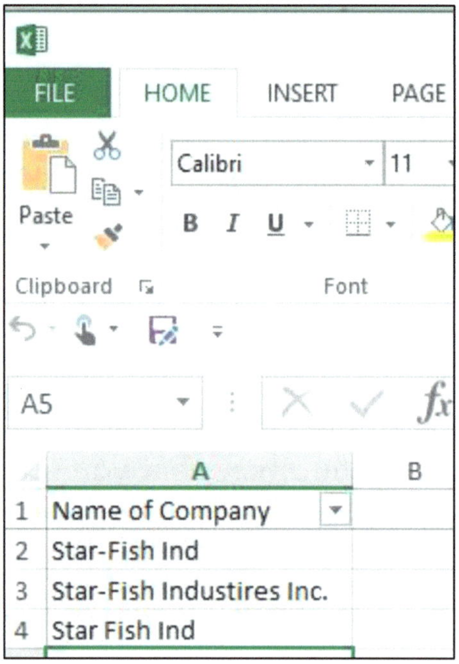

EXAMPLE: CREATING DROP-DOWN LISTS WITH VALID TEXT CHOICES

The following instructions demonstrate how to create drop-down lists with valid text choices.

1. Select the storage column, which is Column K.

2. On the top menu bar, select **Data**, then 

, and then **Data Validation**. The Data Validation dialog box appears.

   Column K will be the data validation *storage column* and will be updated shortly. The storage column holds your data validation drop-down list.

3. In the Allow option, select **Text Length**.

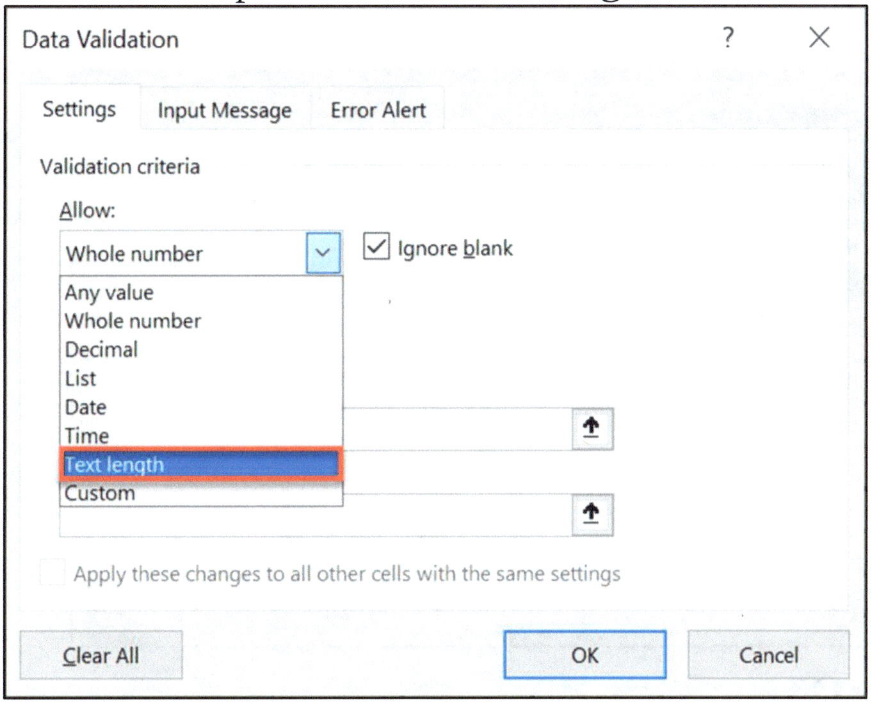

4. Specify the following options under Settings:
Allow: **Text length**
Data: **less than**
Maximum: **25**

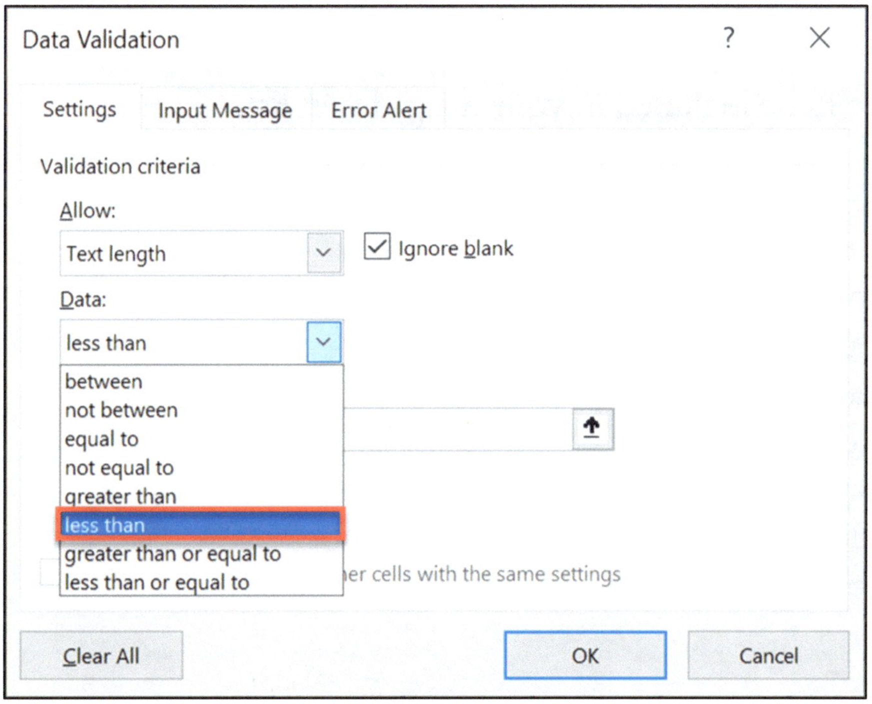

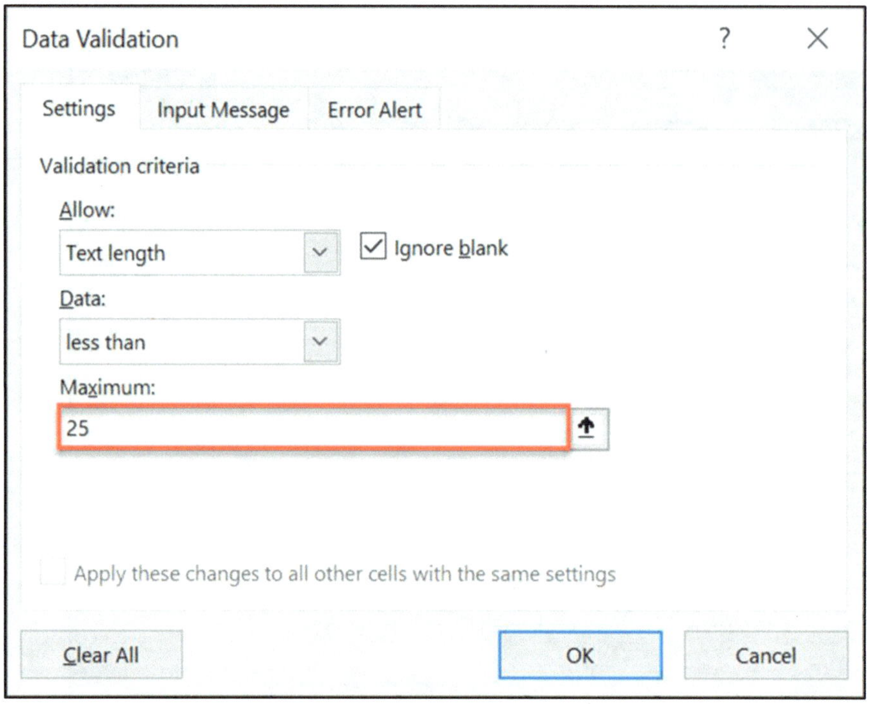

5. Click **OK**.
6. To create a drop-down list, add another column heading called "Species of Mushrooms" in Column K (Cell K1). That is, the same column heading we have for Column D.
7. In the cells directly below the heading, starting with Cell K2, enter the following types of mushrooms, which will become the options that will be included in your drop-down list.

Portabella
White Button
Cremini
Portabello
Shiitake
Oyster
Enoki
Porcini
Morel

Note that you might have to widen Column K for the column heading text to fit. Column K now looks like this.

8. Select Column D, which is where you want the drop-down list to appear.

9. Open the Data Validation dialog box again by clicking  and then **Data Validation**.

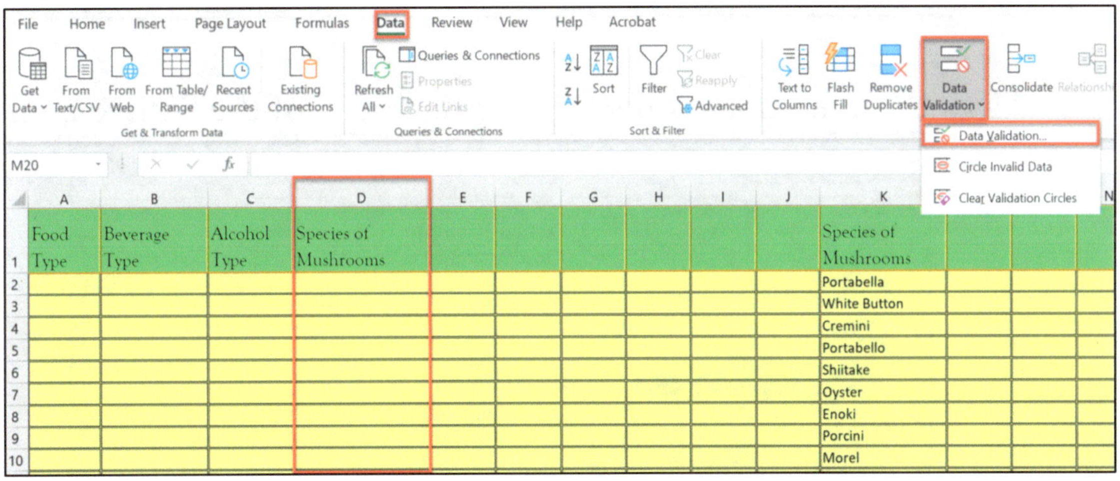

10. In the Allow option, select **List**.

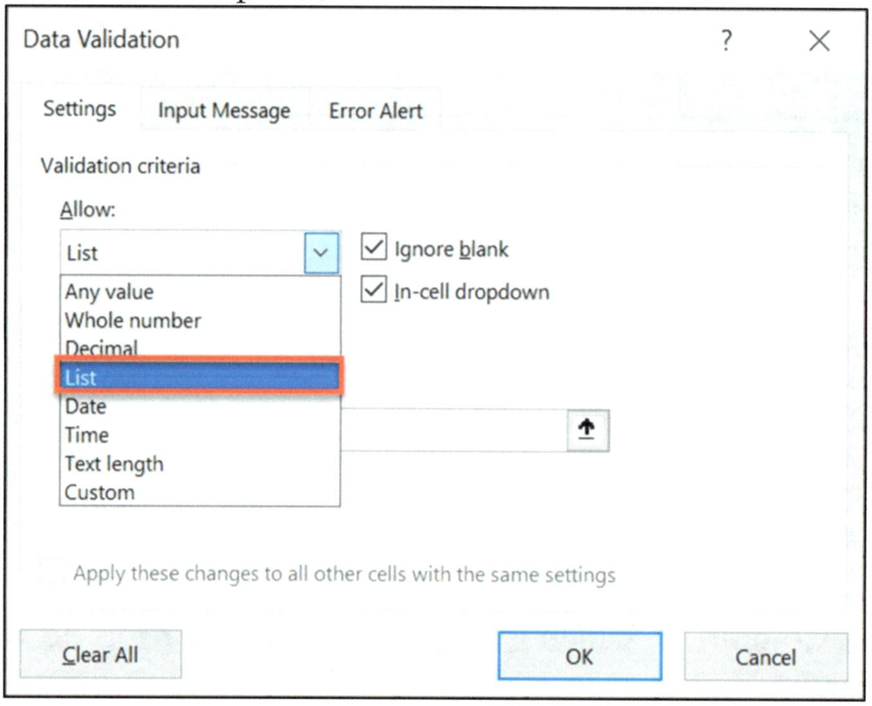

Important! Be careful with the next steps. These steps must be done exactly as written or the data validation will not work.

11. Click the small square on the right side of the Source box.

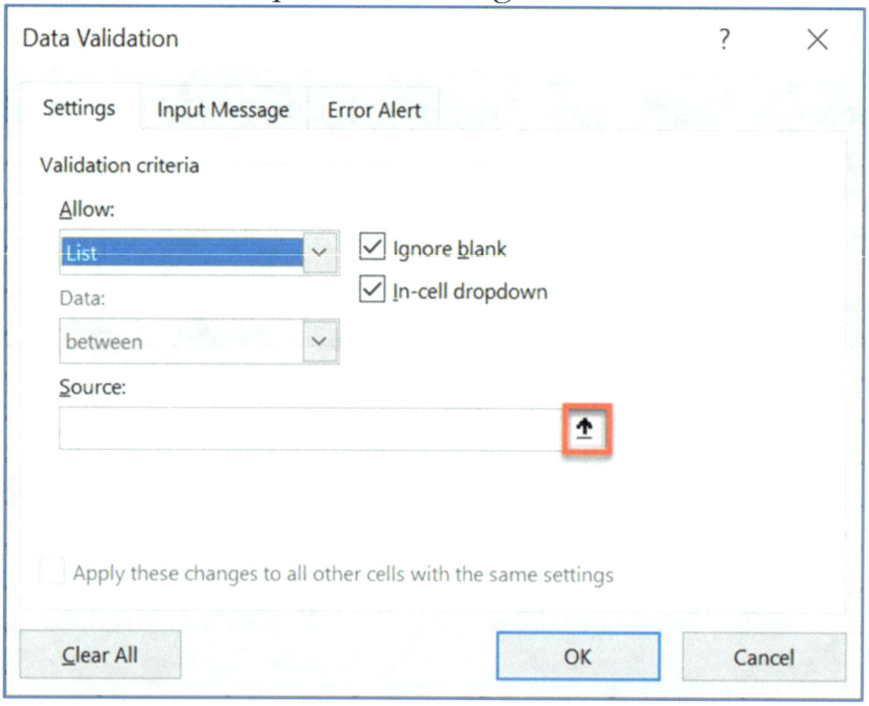

The Data Validation box appears.

12. Click the first cell in the list in Column K, where you have the first type of mushroom that you want to add to the drop-down list. For our example, we want to select Cell K2, "Portabella."
13. Without raising your mouse, drag your cursor to the bottom of the list to select the whole list, that is, to Cell K10, "Morel." You can see that the text you want to add to your drop-down list is selected.
14. Press **Enter**.

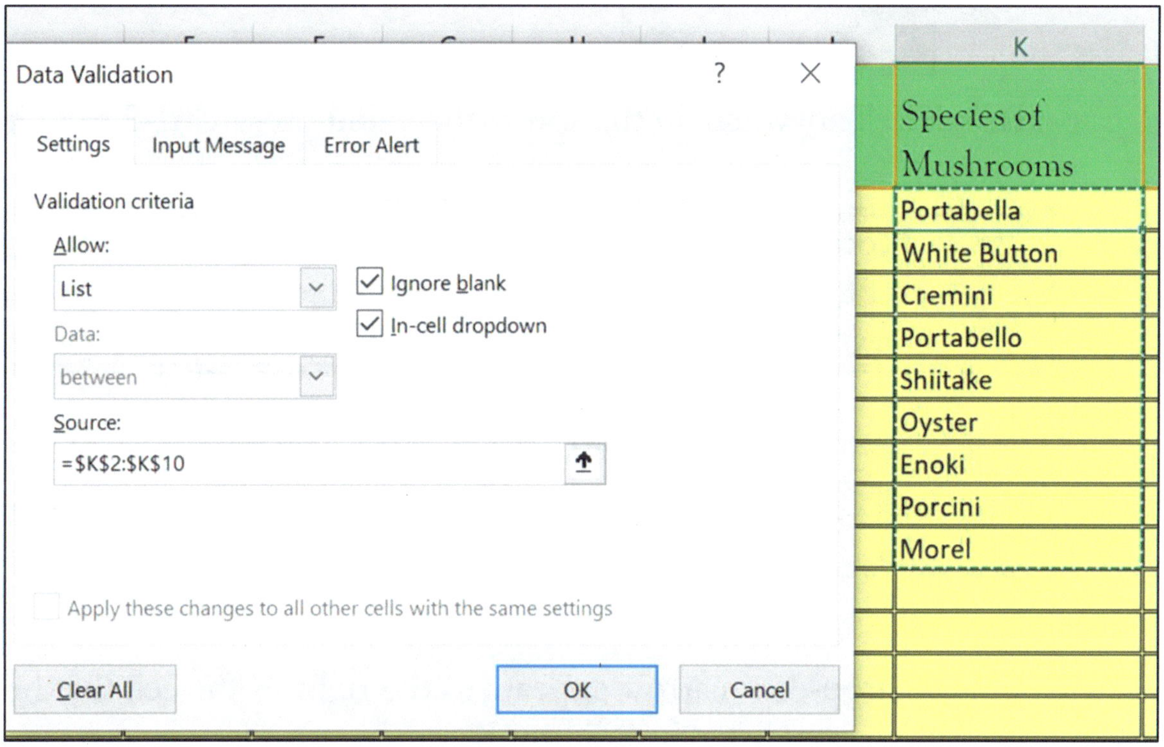

The cell range appears in the Source box in range format, which means that Cells K2 to K10 have been selected.

15. Click **OK**.
    Now the drop-down for Species of Mushrooms has been created. This will minimize typos and incorrect spelling and will both save time and increase accuracy when entering data into the spreadsheet.

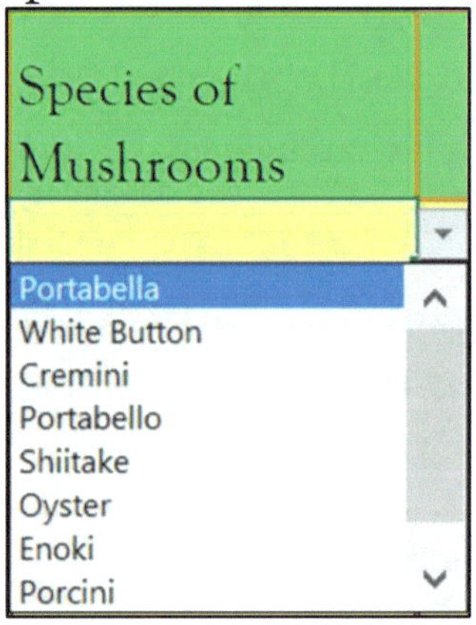

16. Click anywhere in the spreadsheet and press **Ctrl-S** to save it.

Note: Once you create a drop-down list, you can't change its font size or styling. You may be able to do so with programming or another software program used in combination with Excel.

Example: Creating Drop-Down Lists with Valid and Invalid Text Choices

Now let's see what happens when you try to enter text that is not in the data validation list.

1. Click Cell D2.
   A drop-down arrow appears to the right of the cell that has the items you listed in Column K.
2. Click **Portabella** from the drop-down list or type it in Cell D2.
   This is a valid text choice.
3. Type **Tomato** in Cell D3.
   You will see that Excel will not allow this.

4. You then have the option to click Retry and enter a valid item or Cancel. Let's select **Cancel**.

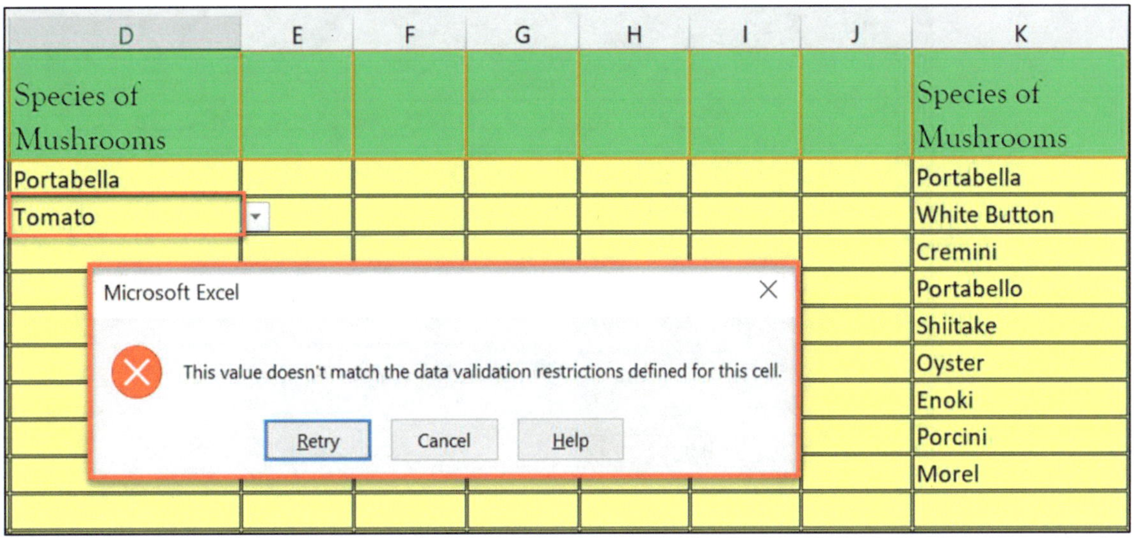

Creating Custom Warning Messages

You can create a customized message that warns you when cells contain data validation.

1. Select the cells to which you want to apply data validation. For this example, select Column B.

2. On the Data menu, click 

, and then select **Data Validation**.

3. In the Data Validation dialog box, select the **Input Message** tab.

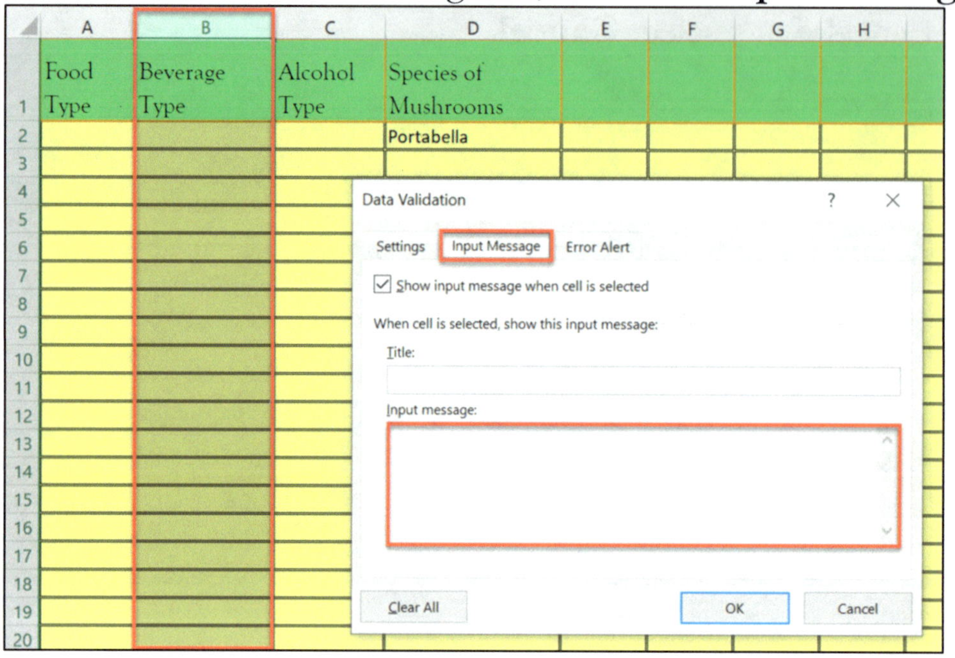

4. Enter the input message you want to see. The message should pertain to the validation rules. For this example, enter the following in the Input message box: "Make sure the SKU code for the Beverage is between 1983 and 3900." Entering something in the Title box is optional. Let's leave it blank for now.

5. Click **OK**.

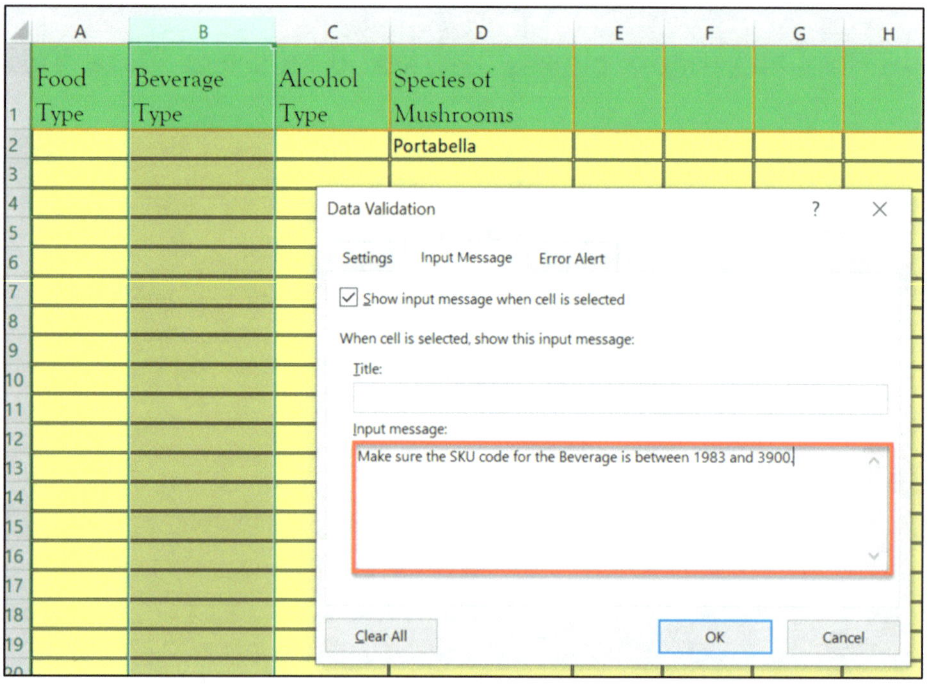

6. When you click on any cell where the data validation has been set up, the input message you created appears.

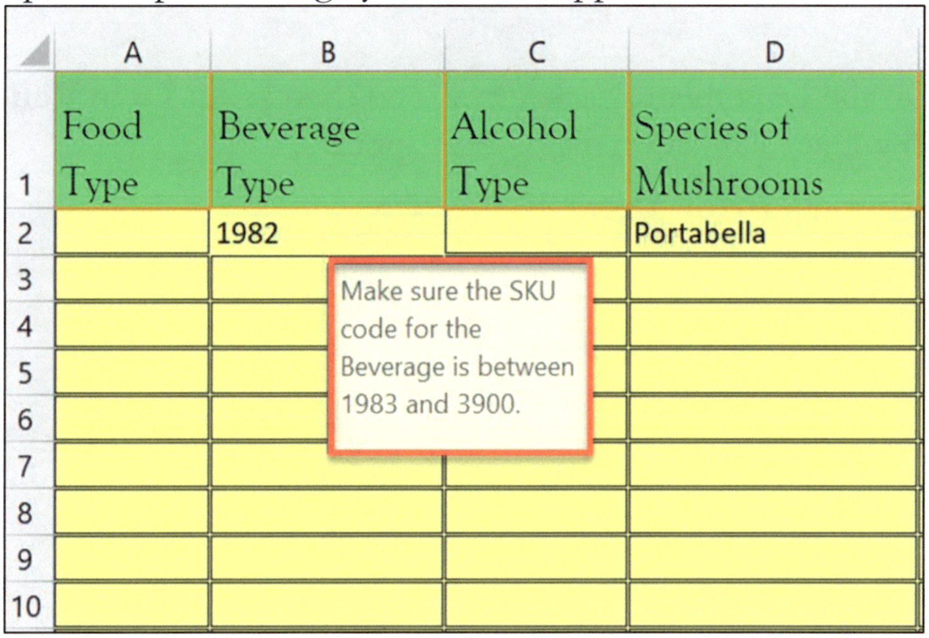

CREATING CUSTOM ERROR ALERTS

You can set up an error alert, which is used when you enter a value that is not within the parameters you set.

There are three types of errors alerts that you can create.

Error Alert Type	Description
Stop	You are not allowed to enter a value that is "out-of-range." This is the default type of warning. You saw this when we previously entered "1982" for the Beverage SKU code, and the value was not accepted.
Warning	When you enter a value out-of-range, you will be required to confirm the value before Excel accepts it.
Information	This alert provides information only. You would receive this message if the value were out-of-range. To accept the value, you would click OK.

1. Select the column to which you want to apply the error alert. For this example, select Column B.
2. On the Data menu, click 
, and then select **Data Validation**. The Data Validation dialog box appears.
3. Select the **Error Alert** tab.

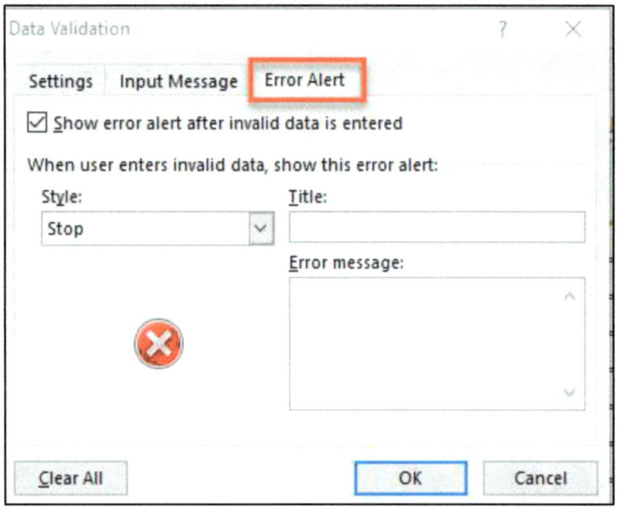

4. Select the **Stop** error alert style.

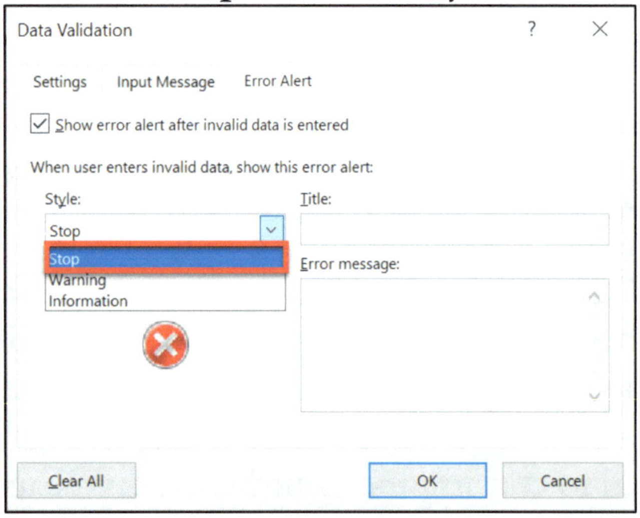

5. Enter the following error message: "The number you entered does not correspond to the parameters of the SKU numbers designated for Beverage Type."

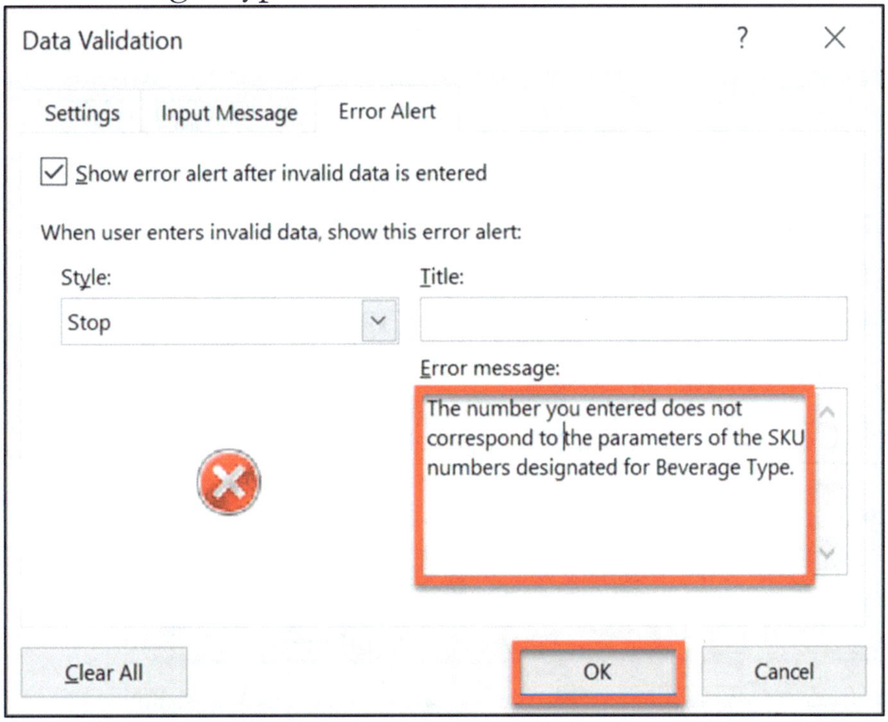

6. Click **OK**.

7. Click anywhere in the spreadsheet and press **Ctrl-S** to save. Enter an out-of-range number, 1966, in Cell B2. The new error alert you created appears. You can click **Retry** to enter a new value or **Cancel**. Let's click **Cancel**.

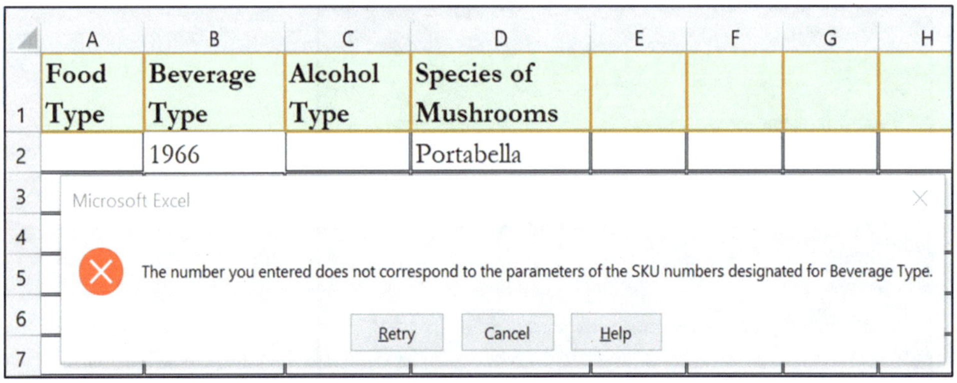

CLEARING DATA VALIDATION FROM THE HEADER ROW

Now I will show you how to change the heading of Column D, "Species of Mushrooms," which was included in the range of cells that have data validation.

Changing the column heading creates an error in data validation, so you must clear the data validation from that cell first.

1. First, let's get the data ready for this example. From the data validation drop-down list in Cell D3, select **Shiitake**. Then in Cell D4, select **White Button**.
2. Click Cell D1, the column heading.

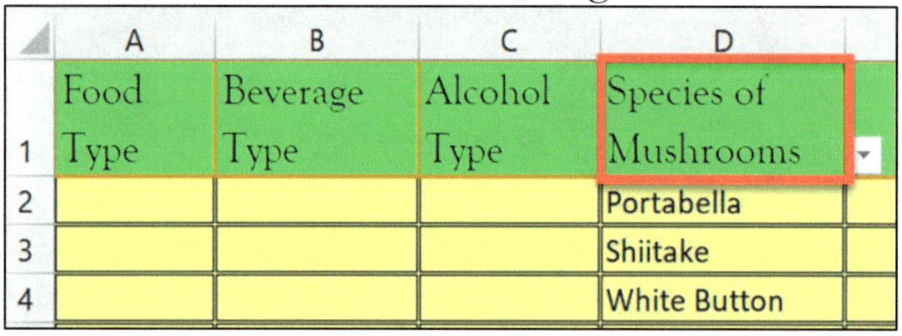

3. On the Data menu, click 

, and then select **Data Validation**. The Data Validation dialog box appears.
4. Select **Clear All**.

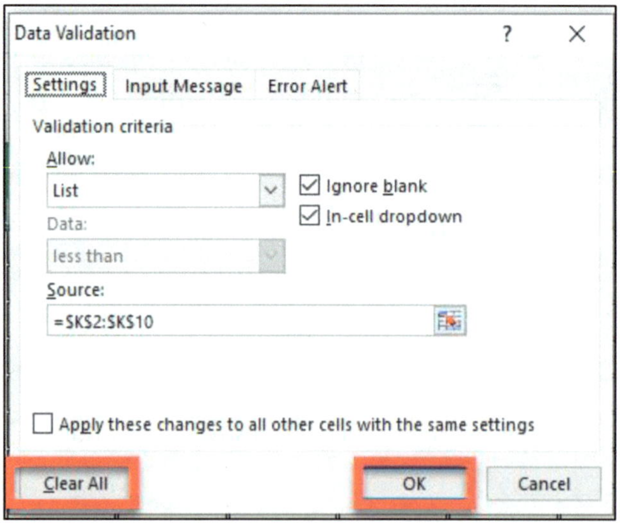

5. Click **OK**.

6. Click Cell D1.
   Species of Mushrooms" appears in the edit box at the top of the screen.
7. Change the cell name in that edit box to **Types of Mushrooms**.

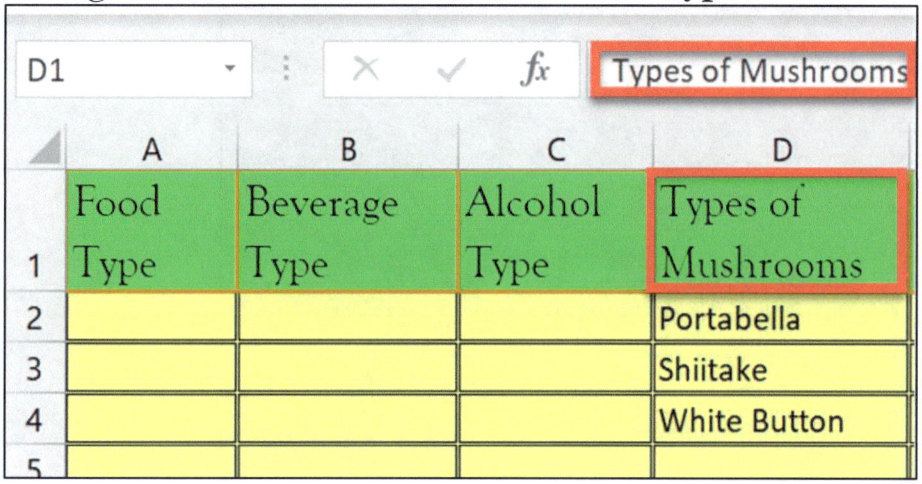

Now you can see that I was able to change the column heading from "Species of Mushrooms" to "Types of Mushrooms."

I only cleared data validation for Cell D1. Data validation is still valid on the rows underneath the header row.

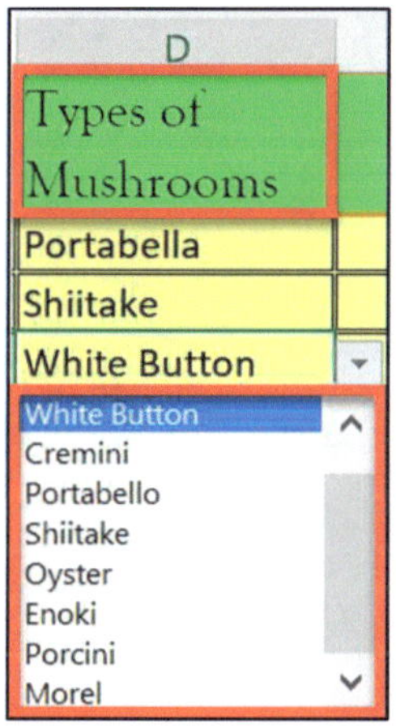

8. Press **Ctrl-S** on your keyboard to save.
9. Click on Cell D4, "White Button," and you can see that the data validation is still in place.

10. Data validation has only been removed from the header row in Column D.

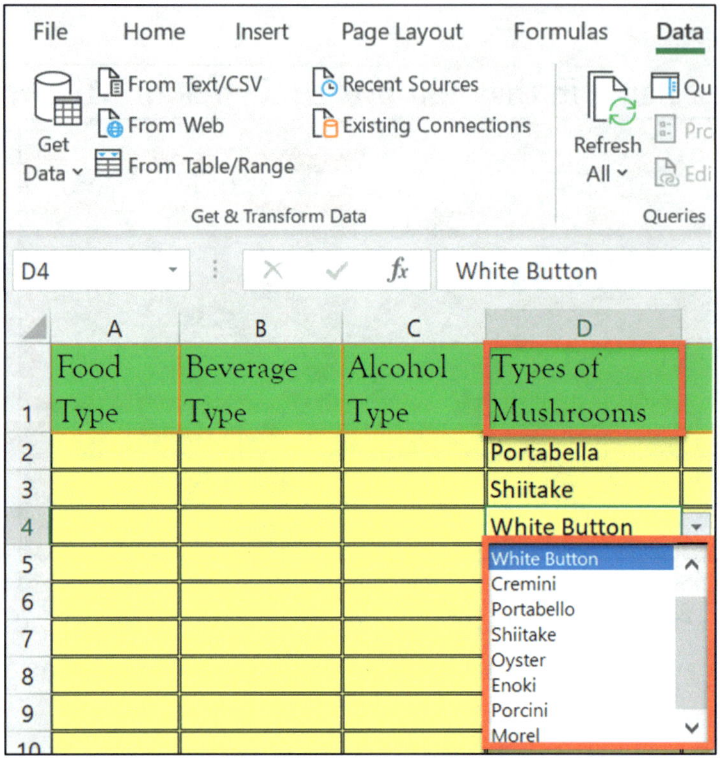

Chapter 3: Filtering and Sorting

This chapter will show you how to:

- Filter spreadsheet data
- Sort in alphabetical order

Filtering Your Spreadsheet Data

Now we want to create drop-down arrows on the spreadsheet.

This is an important Excel feature called filter, which allows you to search for items that meet certain criteria in your list (column).

1. Place your cursor in a cell in the top row, such as "Types of Mushrooms."
2. On the menu bar, select **Data**, then the **Filter** button.

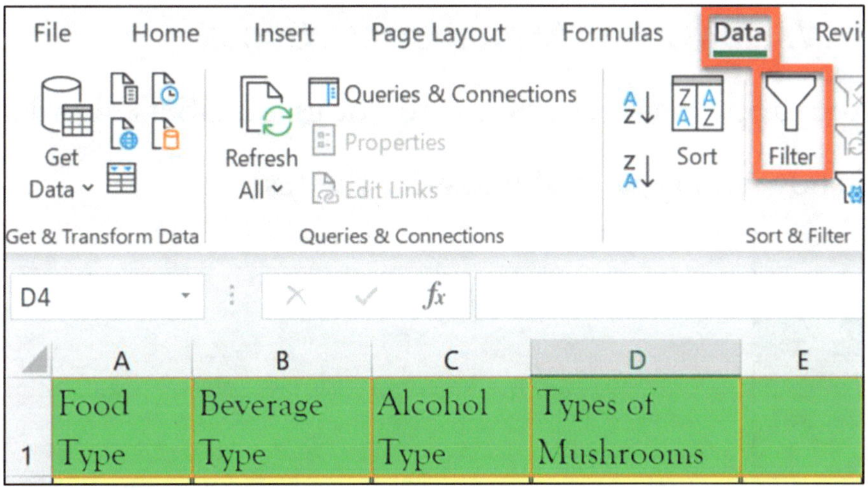

Now you can see that we have created drop-down arrows for "Food Type," "Beverage Type," "Alcohol Type," and "Types of Mushrooms."

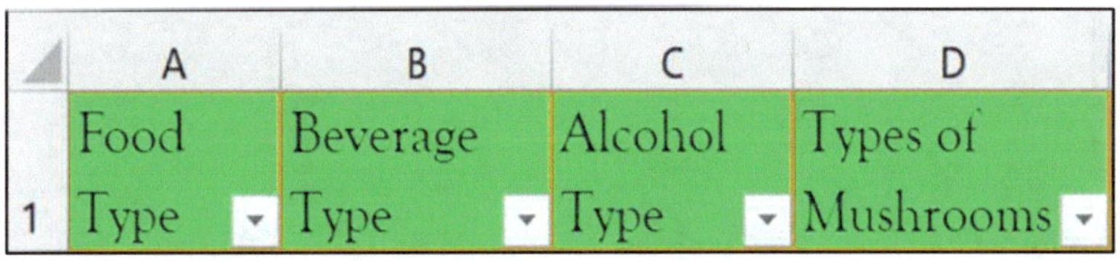

3. When you have data filled into the cells in these columns, you can filter the column for data items that meet certain criteria.

4. The filter options window shown below only appears when you select the drop-down arrow in the header row.

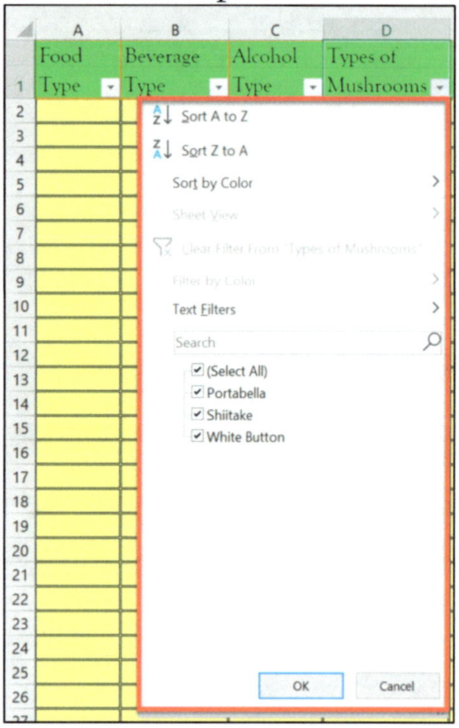

SORTING DATA VALIDATION IN ALPHABETICAL ORDER

Note that the drop-down selections are not in alphabetical order.

Now I will show you how to alphabetize the text.

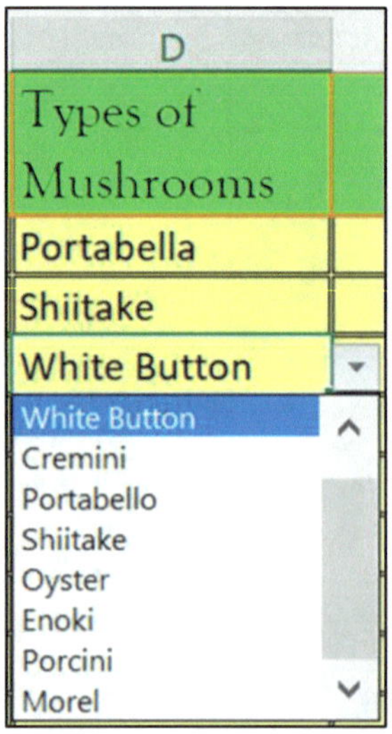

Below is the column (Column K) that we will select to sort in alphabetical order.

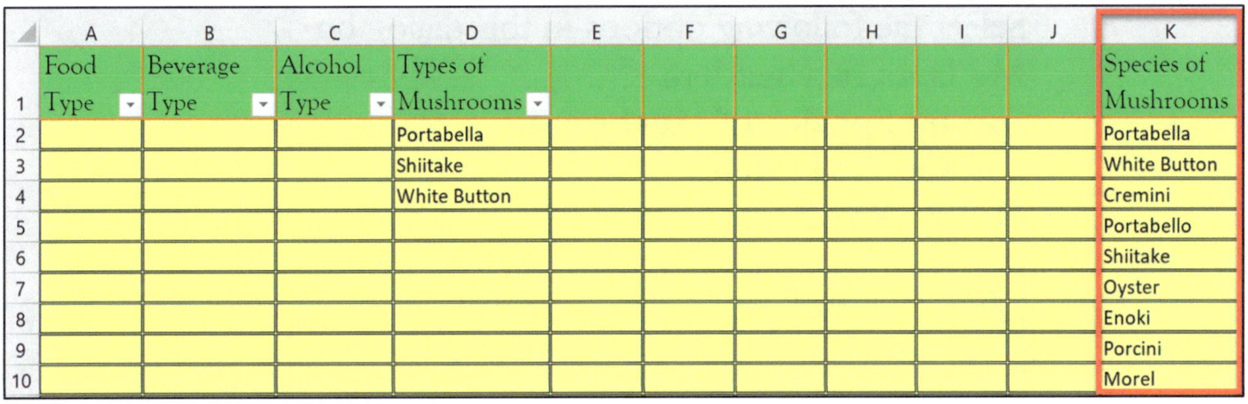

1. Select Column K.

2. In the top menu bar, select the **Data** menu and then click **Sort**.

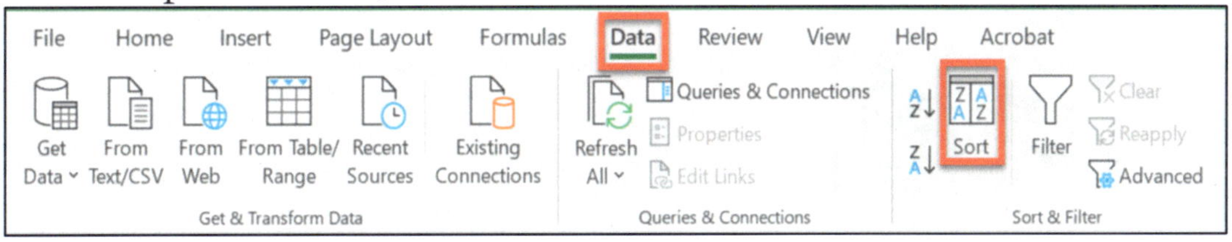

The Sort dialog box appears.

3. Select the following options in this dialog box.
   My Data has headers
   Sort by: **Species of Mushrooms**
   Sort On: **Cell Values**
   Order: **A to Z**

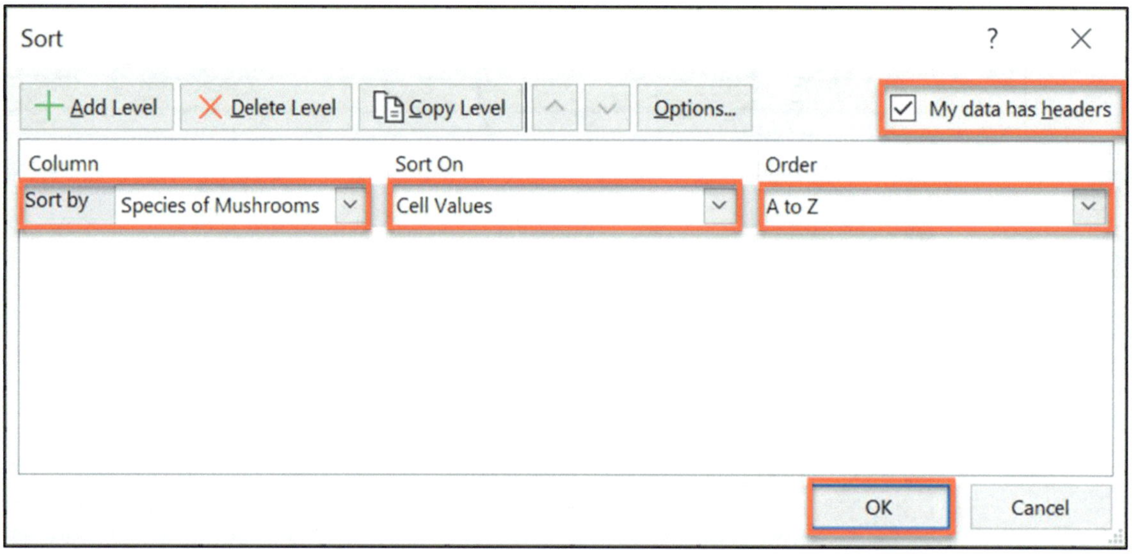

4. Click **OK**.
   Now you can see that the species of mushrooms are in alphabetical order.

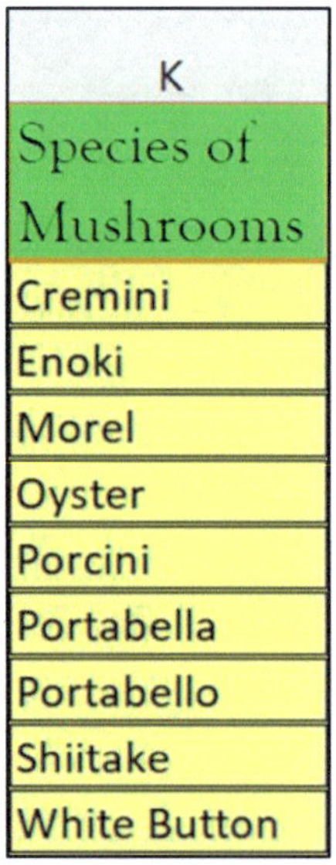

   Because we chose "My data has headers" in the above selection, we kept the header row of "Species of Mushrooms."

5. Click anywhere in the spreadsheet and press **Ctrl-S** to save it.

6. Now when you select the data validation drop-down list, your choices are in alphabetical order.

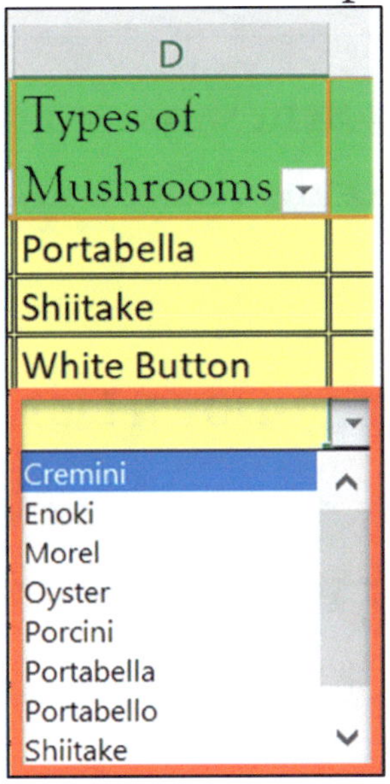

Chapter 4: Adding to the Data Validation List

This chapter contains information on three different ways you can add additional choices to your existing data validation list, after the list has been completed.

Using a Keyboard Shortcut to Select the Drop-Down List

If you don't want to have to press the drop-down button constantly to use your data validation, there are shortcut keys you can use to do this.

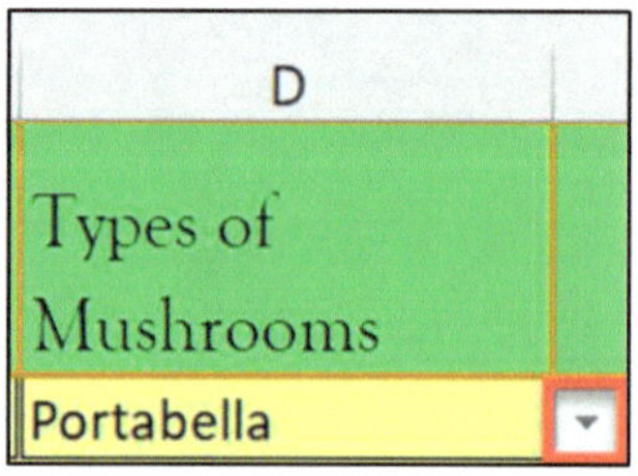

To use the keyboard to open the drop-down list, hold the **Alt** key while pressing the down arrow key.

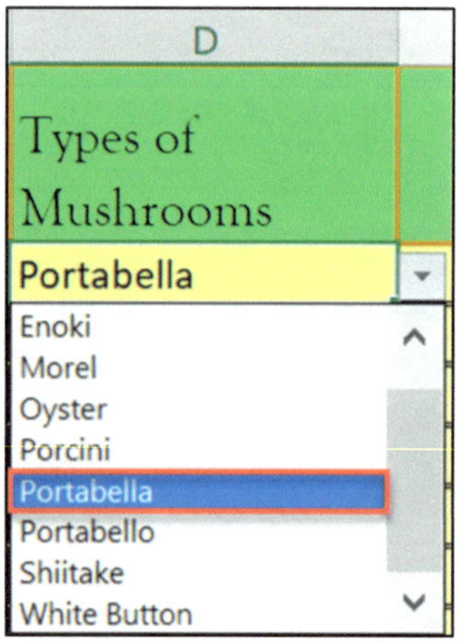

You can then press the down arrow key to select which item you want to select within your data validation. You can move up and down the list using the up and down arrow keys.

ADDING CHOICES TO THE DATA VALIDATION LIST

Now let's add an additional item to the data validation list in Column K.

Let's add a new mushroom type "Agaricus campestris," just below "White Button," in Cell K11.

Now look in the validation drop-down in Column D to see if the new mushroom type appears in the list.

The new addition *does not show up in the list* in Column D. This is because we need to update our validation parameters.

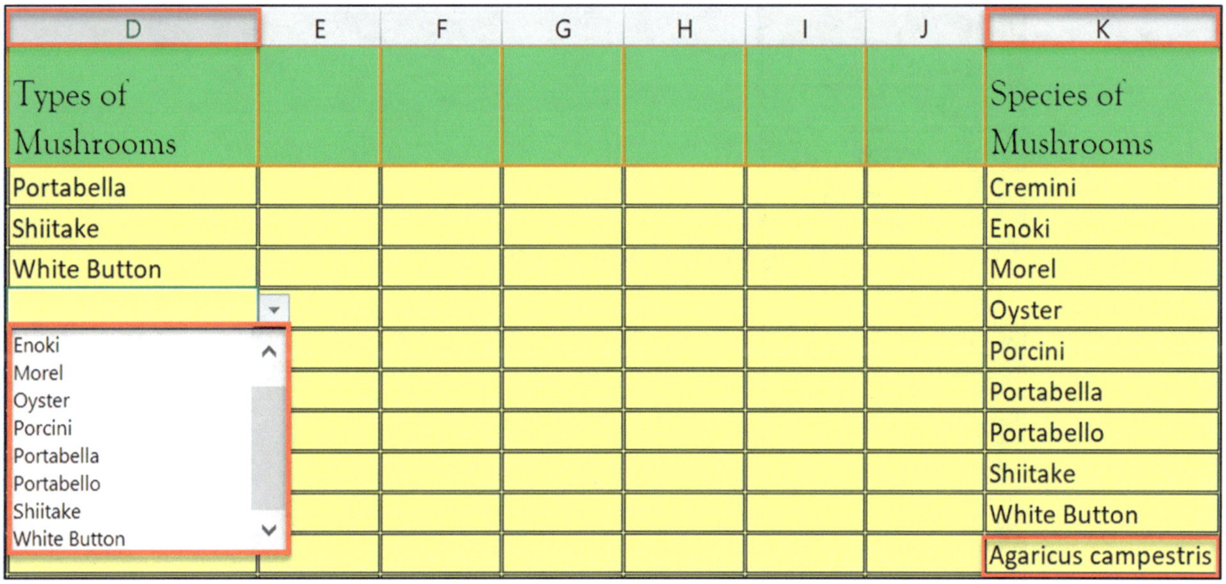

1. Place your cursor in the top-right corner of the cell you want to add to the data validation list. In this case, use Cell K11 with "Agaricus campestris."

The cursor changes to a crosshair.

2. Press the **Shift** key on your keyboard.

Important! You must press the **Shift** key to prevent overwriting the existing cell text.

---

3. Move or drag the item added right below "Morel."

4. Remove your hand from the **Shift** key.
"Agaricus campestris" is now below "Morel."

You can now see that "Agaricus campestris" is in the data validation drop-down list, which is Column D in our example.

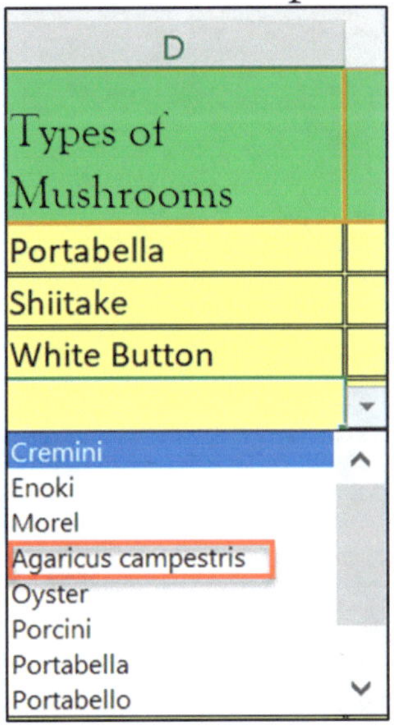

5. Although we are not going to do this for this exercise, if you wanted to sort the list in alphabetical order, see "Sorting Data Validation in Alphabetical Order" on page 46.
6. Press **Ctrl-S** to save.

Adding to the Data Validation List by Inserting a Row

This is my favorite way of adding new items to a data validation list. It is also the easiest way.

1. Place your cursor in Cell K7, "Porcini" and right-click.
2. Select **Insert.**

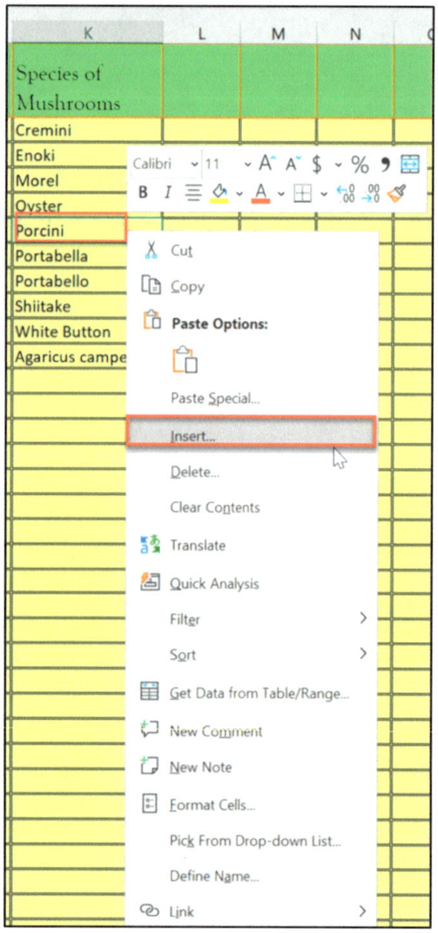

3. Select **Shift cells down.**

4. Click **OK**.

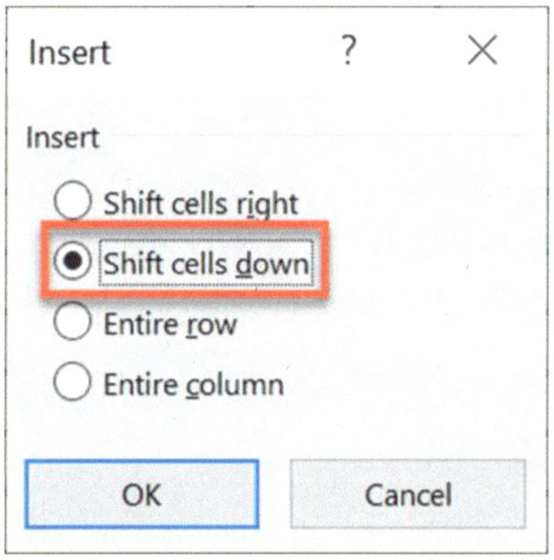

Now you can see that there is a blank cell above "Porcini." This is where you can add the addition data validation item to the list.

5. Let's add "**Beech**" in the blank cell above "Porcini" (Cell K7).

6. Press **Ctrl-S** to save.

7. Now you can see that "Beech" has been added into the data validation drop-down list.

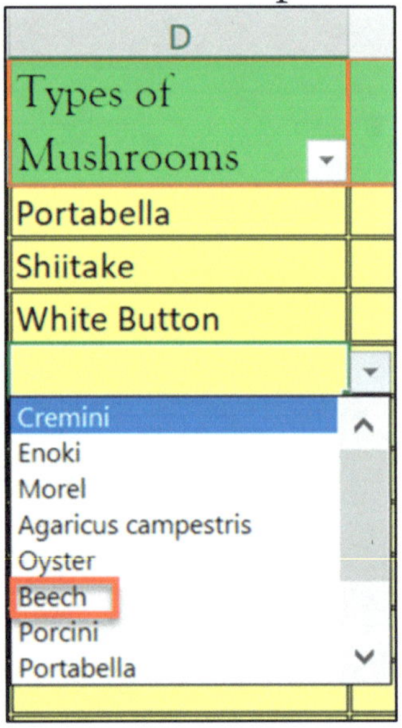

Again, if you now wanted to sort the list in alphabetical order, see "Sorting Data Validation in Alphabetical Order" on page 46.

Adding to the Data Validation List by Creating a Table

This is the third way of adding to a data validation list.

We will create a table that will result in it being easier to make a small addition to a data validation list.

While this takes longer to set up initially, it is easier for future new additions you make to your data validation list.

1. Click anywhere in Column K. For our example, let's click "Agaricus campestris," Cell K5.

   On the top menu bar select **Insert**, then select **Table**.

   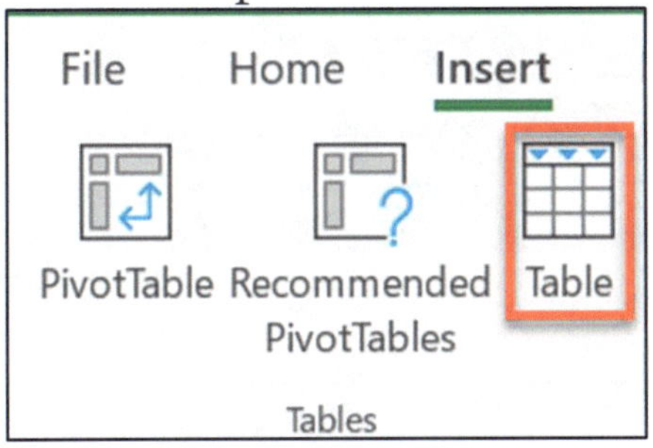

   The Create Table dialog box appears.

Tip: A shortcut for selecting a table is pressing Ctrl-T.

2. If it is not already selected, click **My table has headers** because "Species of Mushrooms" is the table header.
3. Click **OK**.

   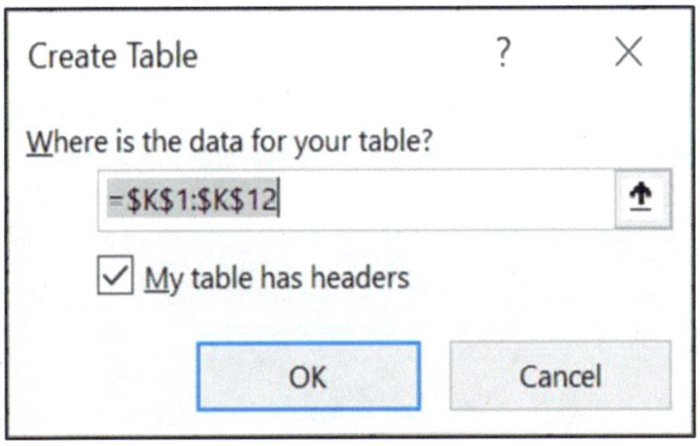

   You can see that I now have a table. Let's clear the formatting.

4. Select the drop-down on the right side of the Table Styles area, which is shown with a red box in the screenshot below.

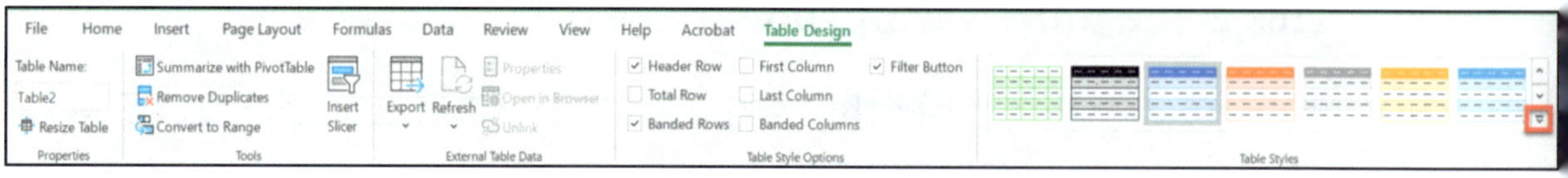

5. Click **Clear** on the bottom to clear the style.

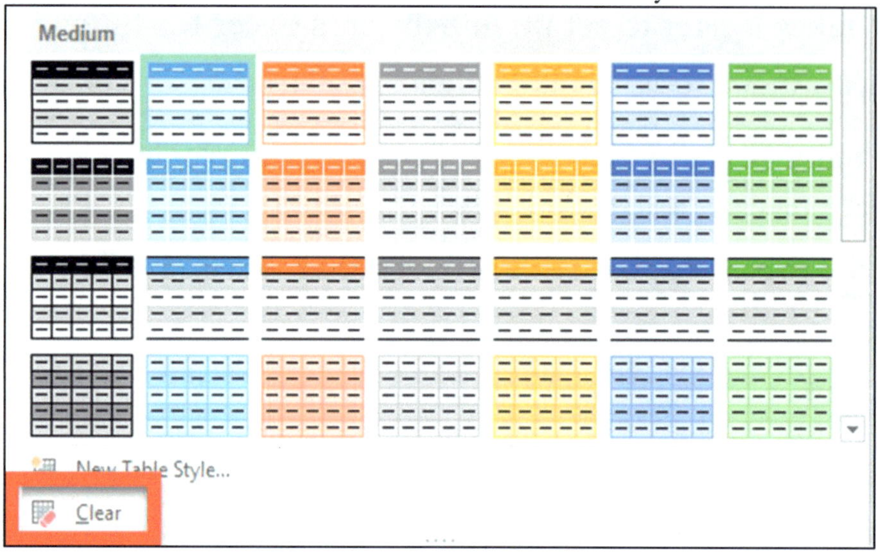

6. Click onto one cell in the data validation drop-down list. (Column K). For our example, let's click on "Agaricus campestris," Cell K5

7. On the Table Design menu, find the "Table Name" area on the left, which now reads "Table1." Change the name of the table to "TableDiv" and press **Enter**.

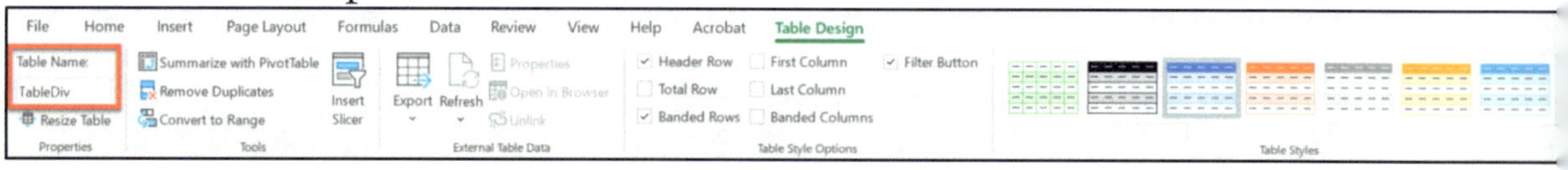

8. Select Column K. Do **not** select the column heading, "Species of Mushrooms," just the list below it (that is, starting with Cremini and ending with White Button).

9. On the top menu bar select the **Formulas** menu, then **Name Manager**.

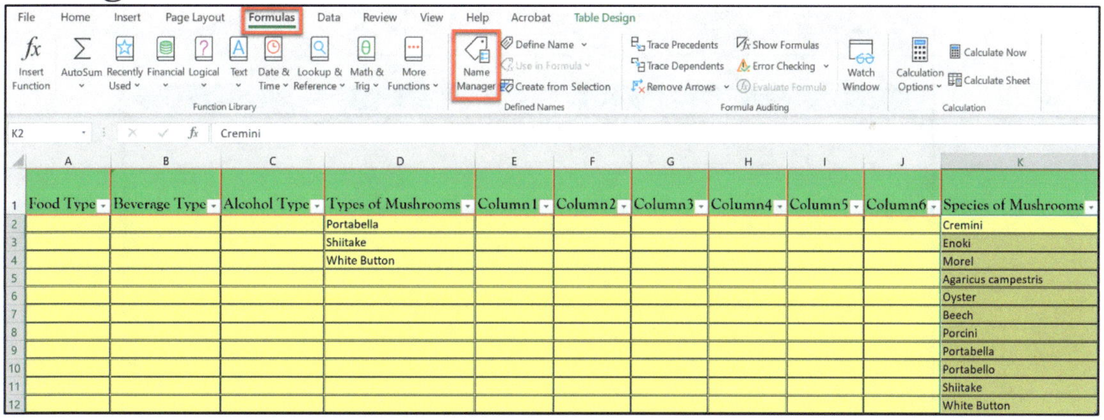

Now you can see the Name Manager dialog box with TableDiv that we created.

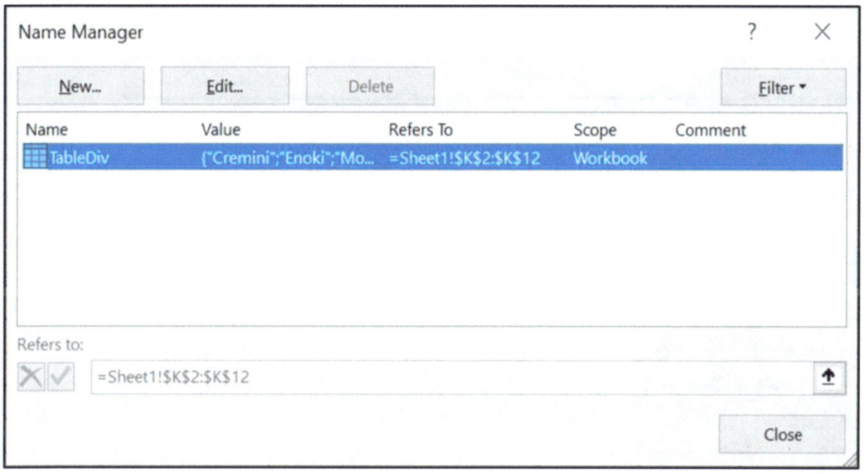

10. Click **New**.
This creates a name range, which is a collection of selected cells. The New Name dialog box appears.

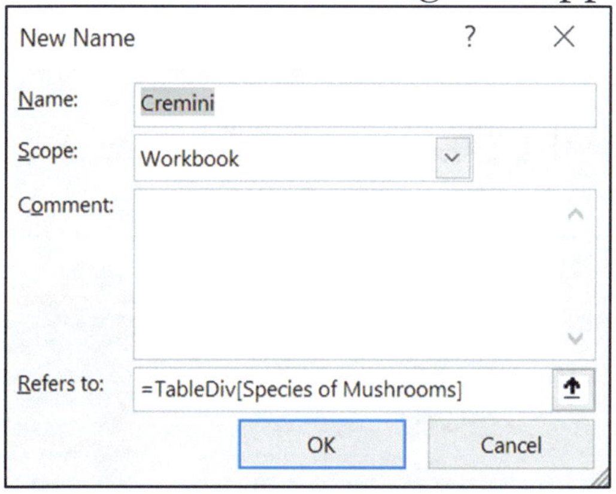

11. Change the name from "Cremini" to "Mushrooms," then click **OK**.

12. Now you can see my new name of "Mushrooms."

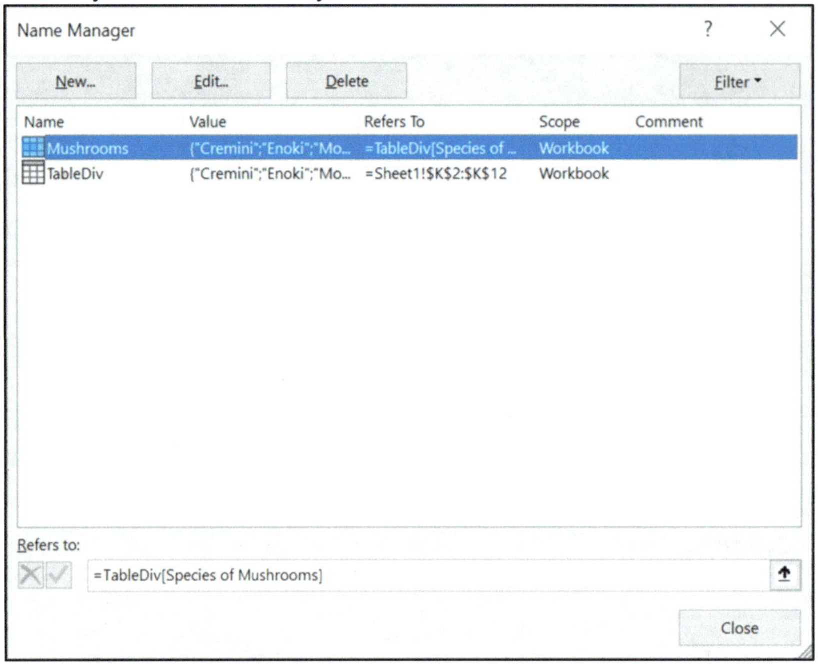

13. Click **Close**.

14. Select Column D, the data validation list.

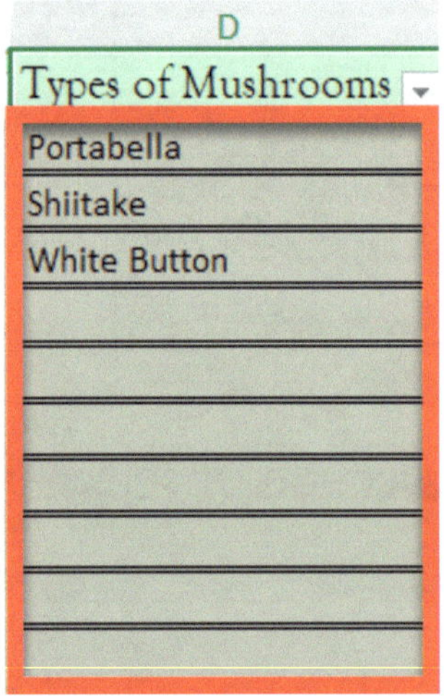

15. On the top menu bar, select the **Data** menu, then **Data Validation**.

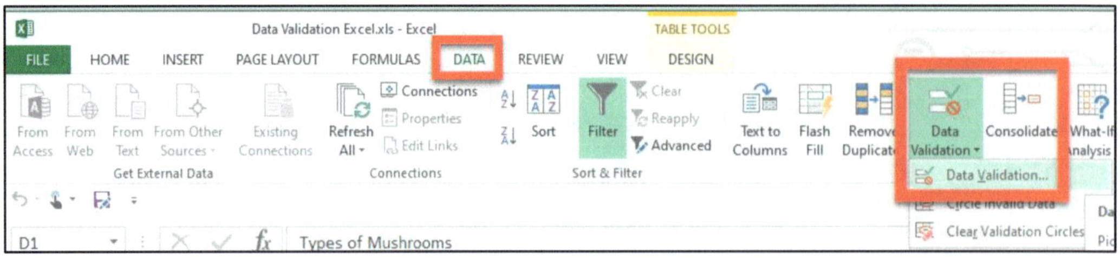

The Data Validation dialog box appears.

16. On the Settings tab, change the Allow field to **List**.
17. Click **OK**.

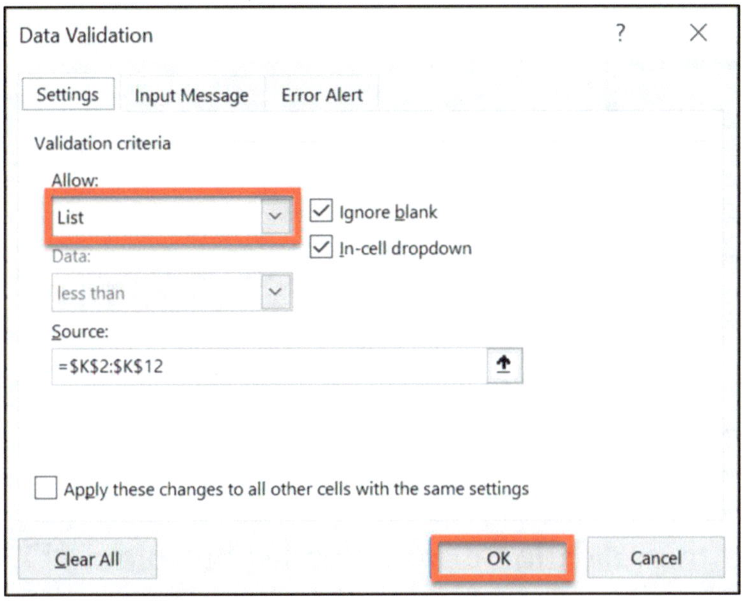

18. Press **Ctrl-S** to save.

19. Let's add a new species of mushrooms to the data validation storage list. In Cell K13, under White Button, type in **Maitake**.

20. If you look at the validation drop-down list in Column D, you can see that the new type of mushroom (Maitake) shows up in the drop-down list.

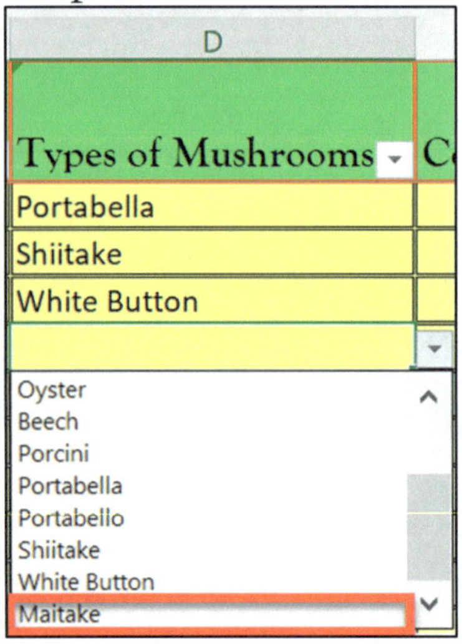

21. Press **Ctrl-S** to save the spreadsheet.

Now that you have created this table each time you want to make an addition to the data validation, simply type your addition into the bottom of column "K" and it will appear in the data validation.

Chapter 5: Changing the Storage Column Information

This chapter will show you a few different ways to move your data validation storage column information

Sometimes it is more convenient to have the information in a location that will not clutter your spreadsheet.

Some options on what you can do with your data validation column follow.

Moving the Storage Column

You may be wondering what to do with the data validation storage column.

The column might start to bother you or just make your spreadsheet too busy.

Remember that Column K is the current location of your data validation list.

One way you can simplify your spreadsheet is by moving the storage column to another sheet in the same workbook.

1. Select Column K, the data validation storage column.
2. Right-click on the column.
3. Then select **Cut** (a shortcut is to press **Ctrl-X**).

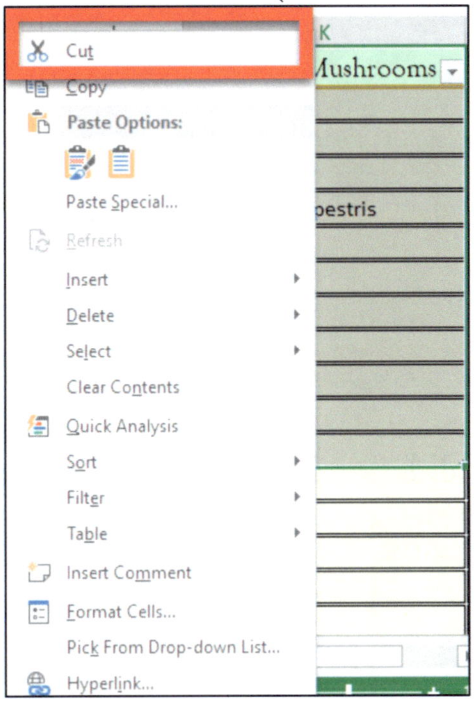

4. Create another sheet in this same workbook by clicking the plus sign to the right of Sheet1 at the bottom of the spreadsheet.

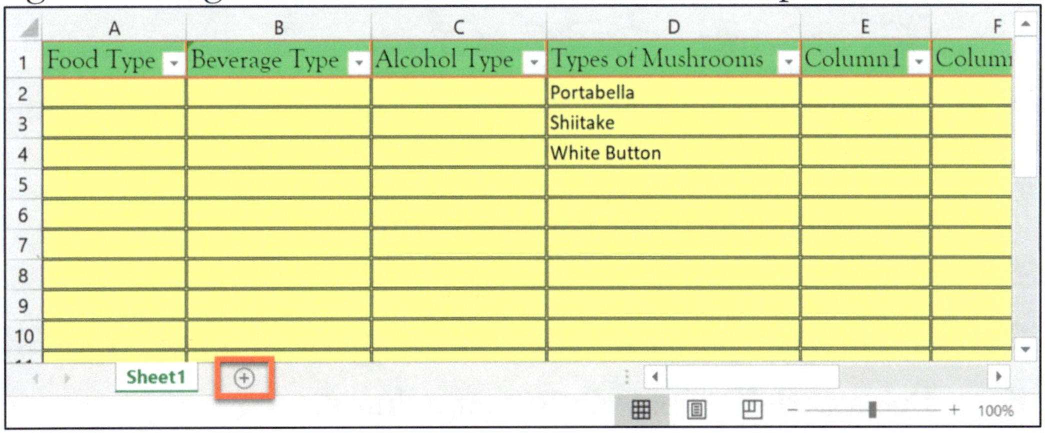

This creates Sheet2.

5. Right-click the words "Sheet2" and then select **Rename**.

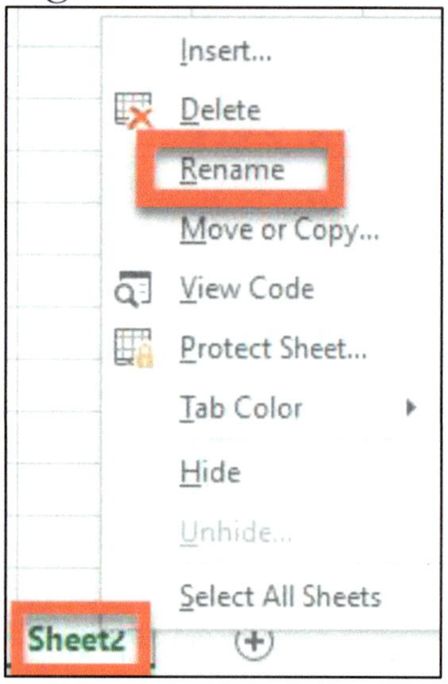

.

6. Change the sheet name to "Validation Back-Up" and press **Enter**.

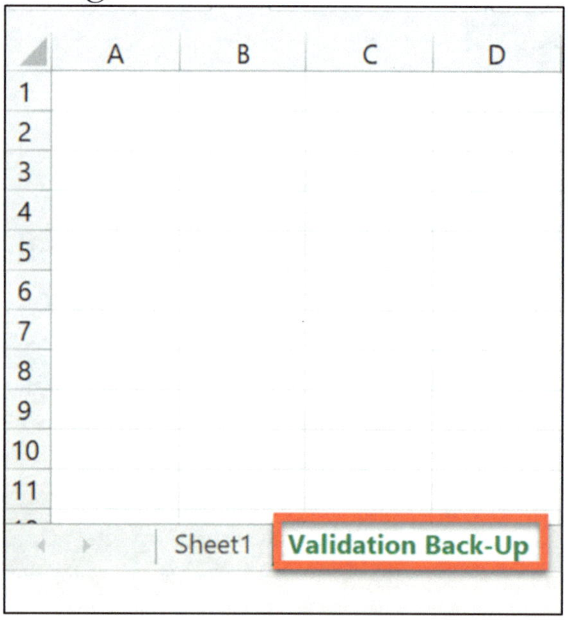

7. On the Validation Back-Up sheet, place your cursor in Cell K1.
8. Press **Ctrl-V** to paste it in that column.

9. Press **Ctrl-S** to save the spreadsheet.
   Now going back to the original Excel spreadsheet, "Sheet1," you can see that the data validation parameters are still in place.

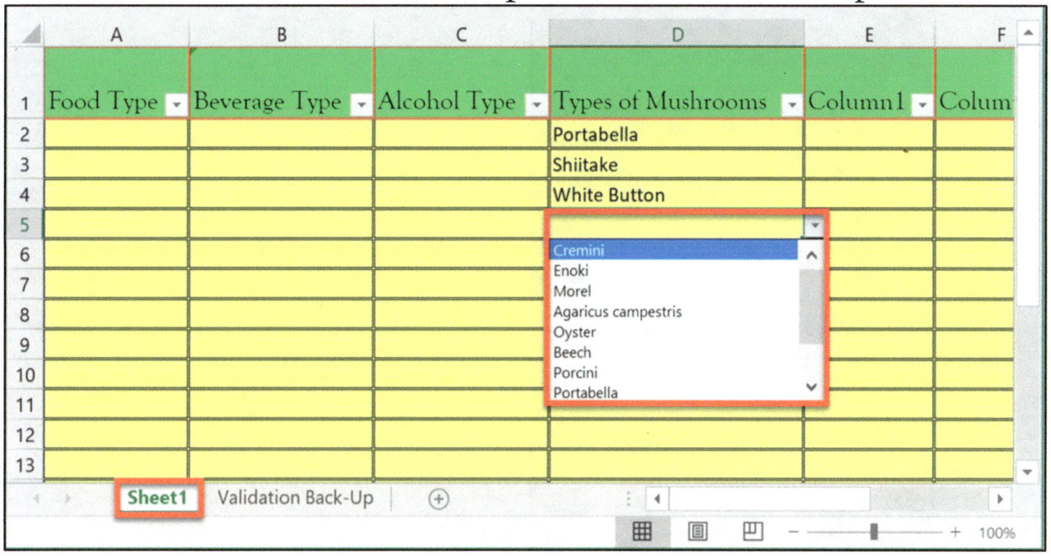

Hiding the Storage Column

Here is another way to make the validation list invisible on your working spreadsheet. First let's move the list from the "Validation Back-Up" sheet back to your main sheet, Sheet1. Once you have done that, follow the steps below.

1. Select Column K, the storage column.
2. Right-click on the column.

3. Select **Hide**.

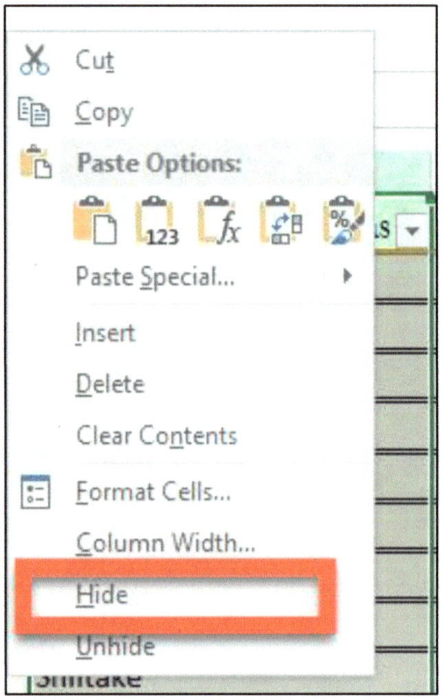

4. Now you see that the column has been hidden.
   Now let's unhide it so that we don't forget it is there.

5. Right-click after you have highlighted both the columns to the left and right of the hidden column (in this case, highlight Columns J and L). ***You can also place your cursor at the corner between two columns.***

6. Then select **Unhide**.

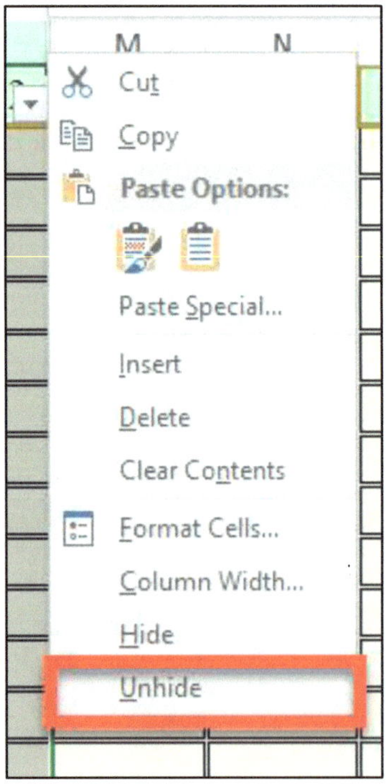

Now you can see that the hidden column has returned.

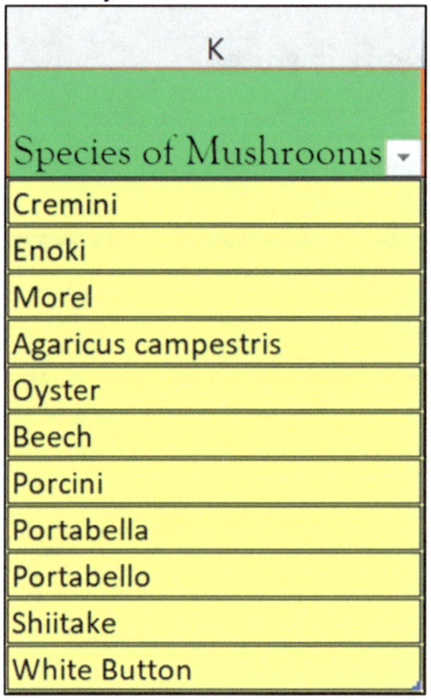

K
Species of Mushrooms
Cremini
Enoki
Morel
Agaricus campestris
Oyster
Beech
Porcini
Portabella
Portabello
Shiitake
White Button

HIDING THE STORAGE COLUMN USING THE TOOLBAR

1. Click one of the cells in the storage column, which in our example is Column K.
2. On the Home tab, select **Format**.

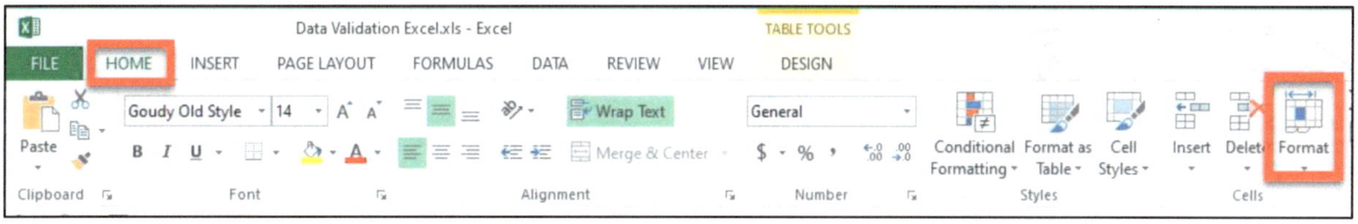

3. On the Format tab, select **Hide & Unhide**.
4. Select **Hide Columns**.

UNHIDING THE STORAGE COLUMN

Now let's unhide it so that we don't forget it is there.

1. Select Columns J and L.

Tip: *Instead of selecting the two columns, you can place your cursor at the corner between the two columns.*

2. Right-click and select **Unhide**.

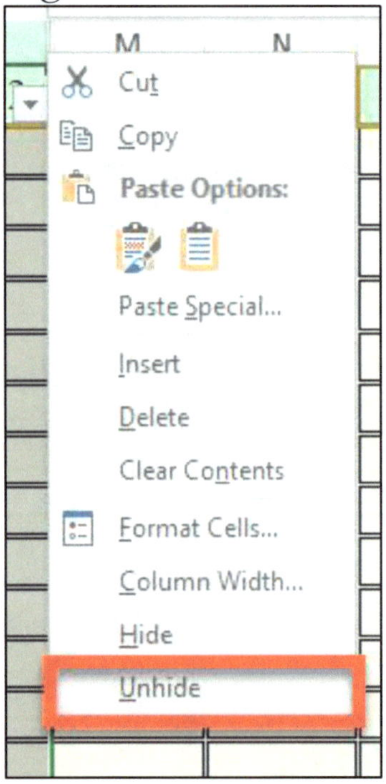

Now you can see that the hidden column has returned.

K
Species of Mushrooms
Cremini
Enoki
Morel
Agaricus campestris
Oyster
Beech
Porcini
Portabella
Portabello
Shiitake
White Button
Maitake

CHAPTER 6: REMOVING DATA VALIDATION

In this chapter, you will learn how to

- Remove data validation from one cell so that you can change only that cell of the column where you have your data validation
- Remove data validation from an entire column

REMOVING DATA VALIDATION IN ONE CELL

This exercise will only remove data validation from one cell.

1. Click Cell D5, the empty cell below "White Button."

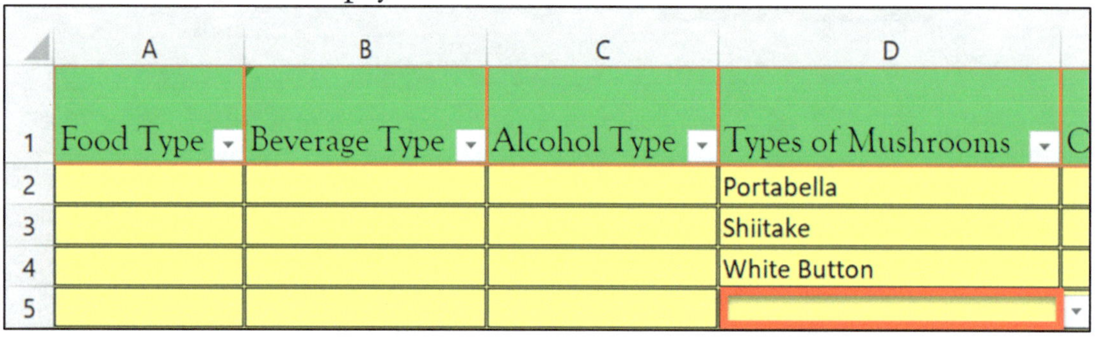

2. On the top menu bar, select the **Data** menu and then **Data Validation.**

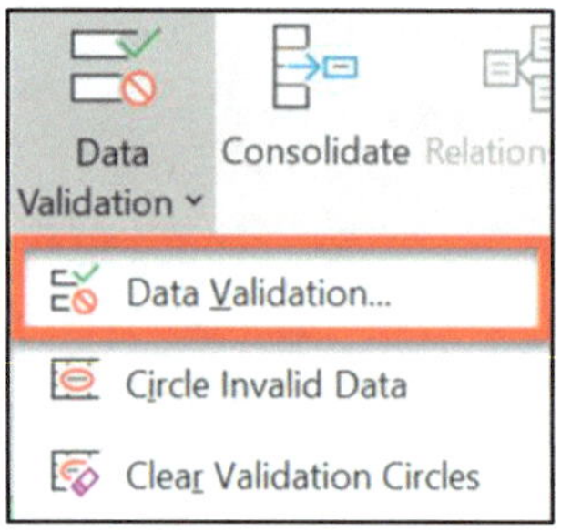

3. Click **Clear All.**

4. Click **OK**.

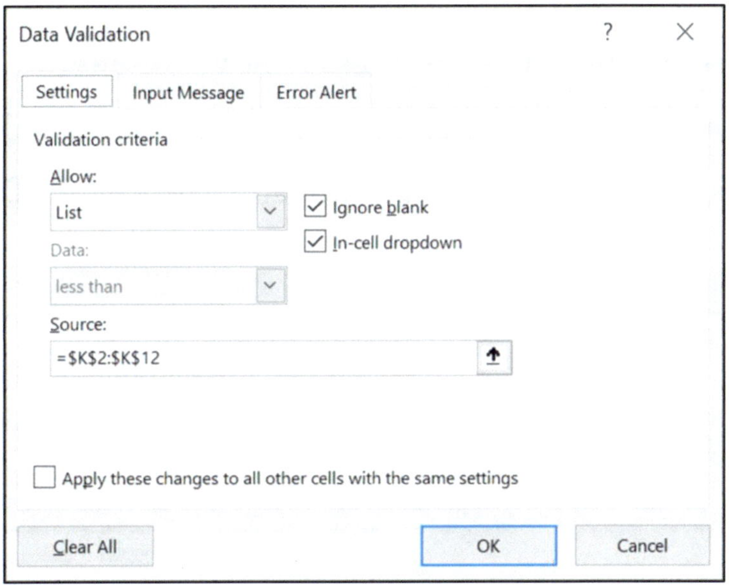

Now you can see that the data validation has only been ***removed from one cell.***

Cell D5 does not have any data validation.

Type in "Pleurotus" in Cell D5 and you can see that it will be allowed even though Pleurotus is not on the validation drop-down list.

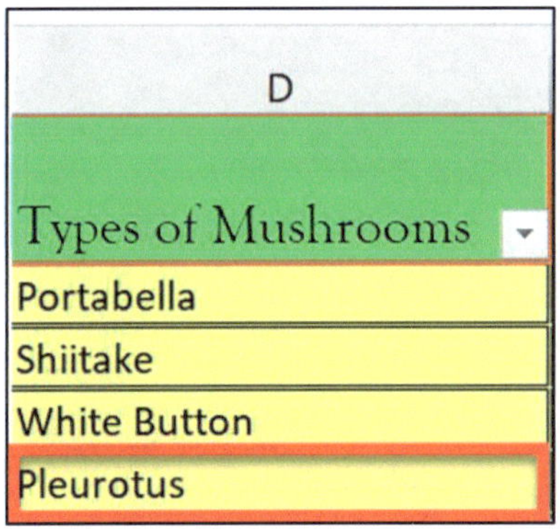

The other cells in that column still have data validation.

REMOVING DATA VALIDATION FROM ALL CELLS

This exercise will remove data validation in an entire column.

1. Select Column D, where we want to remove the data validation.
2. On the top menu bar, select the **Data** menu, then select **Data Validation**.
   A message appears.

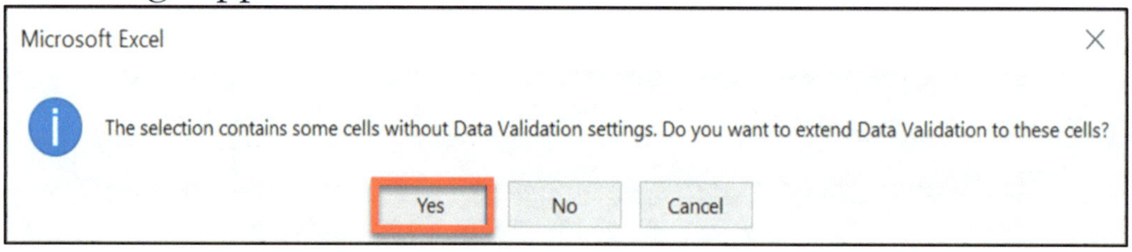

3. Click **Yes**.
   The Data Validation dialog box appears.
4. Click **Clear All**.
5. Click **OK**.

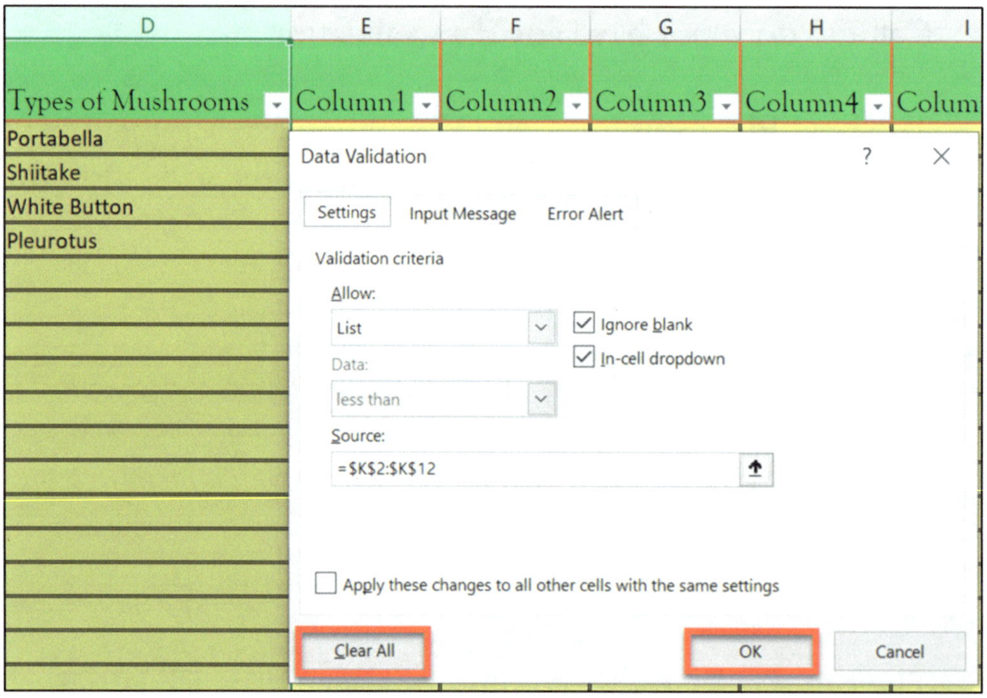

Now you can see that when you click on the cells, the drop-down list for the data validation is gone from the entire column.

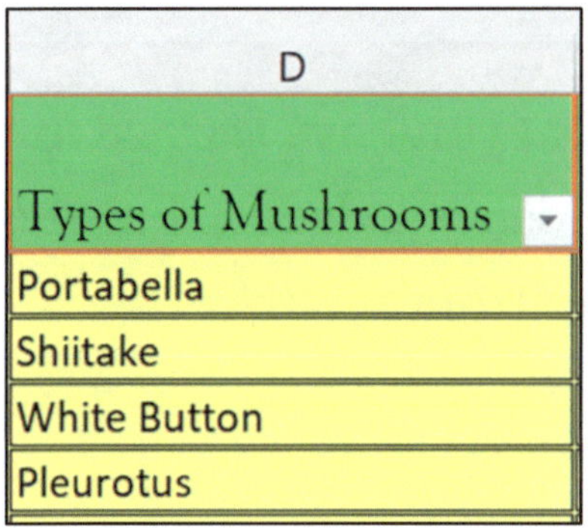

Chapter 7: Applying What You Have Learned to Addresses

This chapter shows you how to apply what you have learned in this book to create complete addresses using data validation. For this chapter, you will create a new spreadsheet from scratch.

Creating the Spreadsheet with Addresses

Excel is great for organizing addresses.

1. Open a new blank spreadsheet.
2. Enter the following information in Cells A1, B1, and C1.
   A1: **Manufacturing Address**
   B1: **Product**
   C1: **Price**
   Make all column headings bold and expand the column so that the text fits without wrapping.

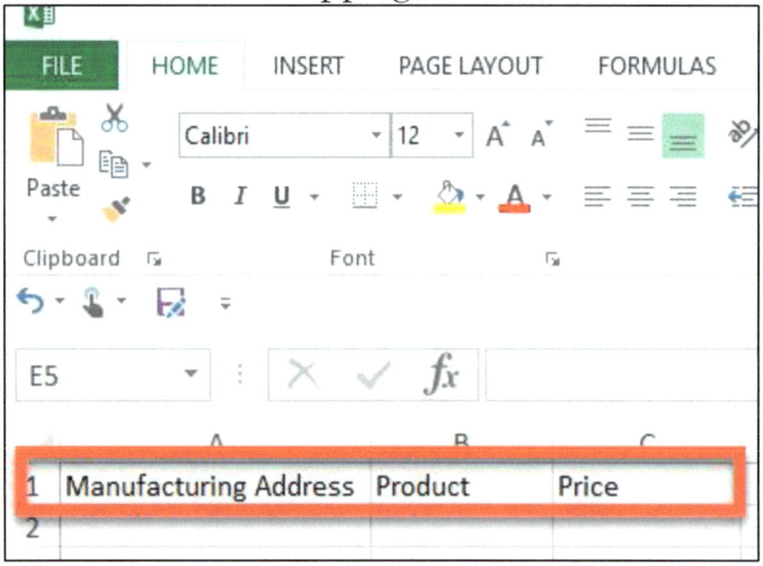

3. Now skip to the first row of Column K and enter **Manufacturing addresses**. Expand the column so that the text fits in Cell K1. This column will be where you store the data that will become the drop-down list for data validation.

4. In Cell K2, enter the entire address even though it spills into other cells: Ted Kay Products 2534 Kennedy Lane Louisville, KY 40202 President: Donald Alda

5. Select Row 1.

   On the top menu bar, select the **Data** tab, then click **Filter**.

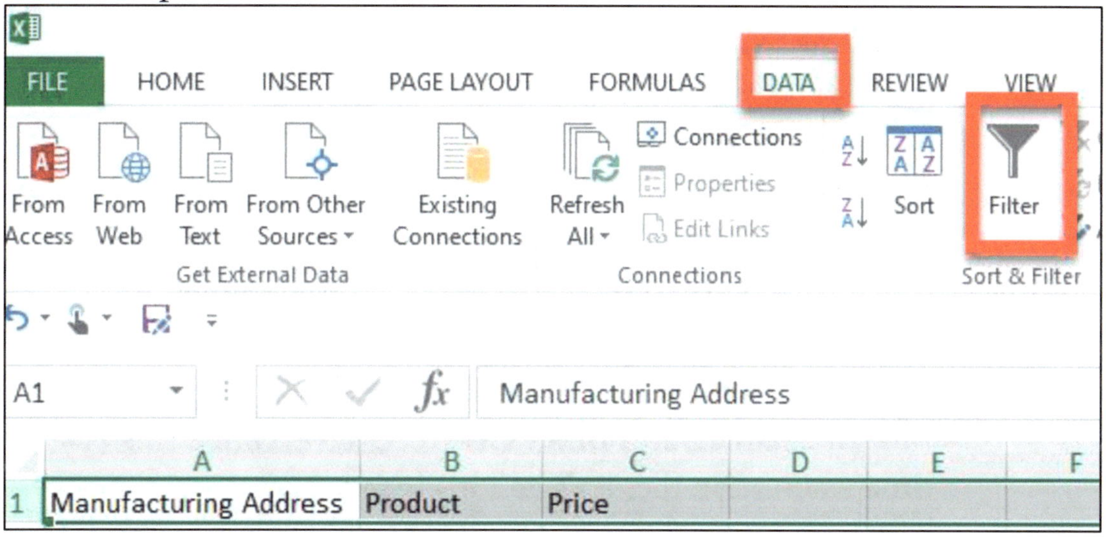

   This creates the arrows so that you can filter Columns A, B, and C.

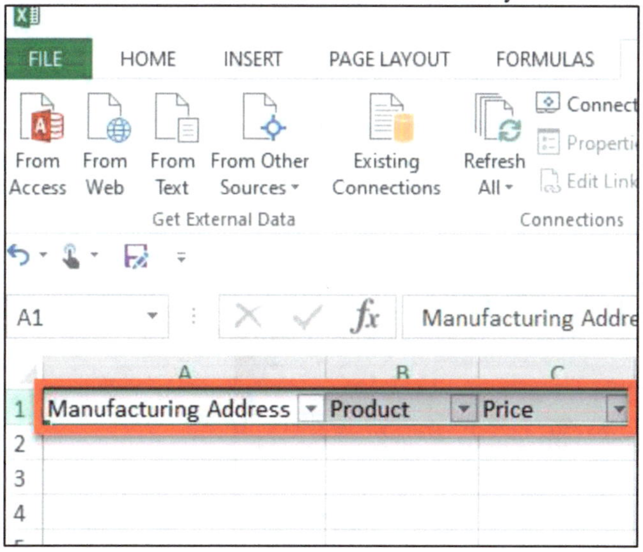

6. On the top menu bar, select the **View** menu.

7. Click **Freeze Panes.**

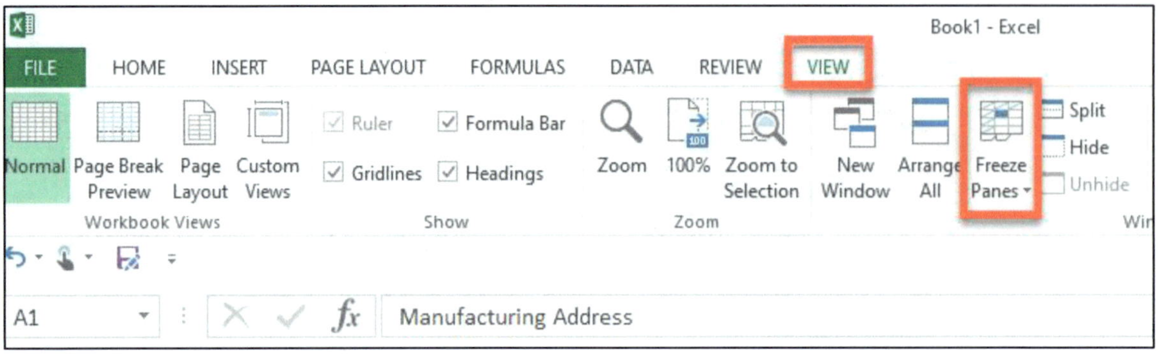

8. Click **Freeze Top Row.**

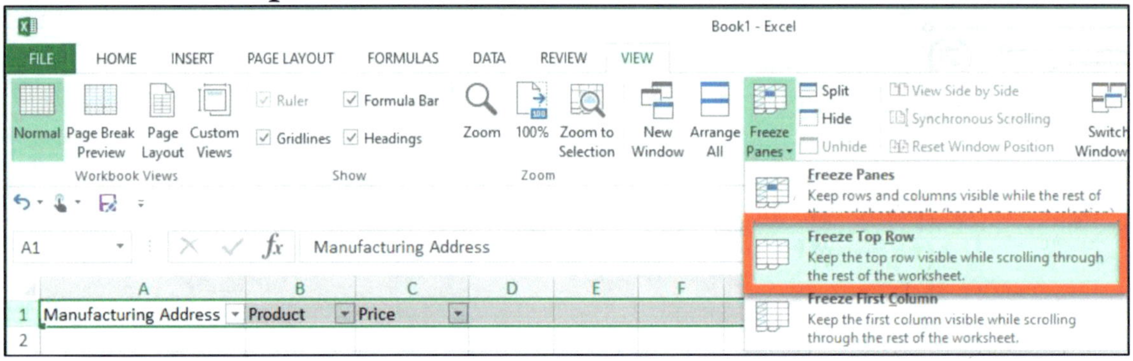

9. This freezes the header row of your sheet. Then click anywhere in the spreadsheet and press **Ctrl-S** to save. Since this is your first time saving the file, let's call it *Address List.*

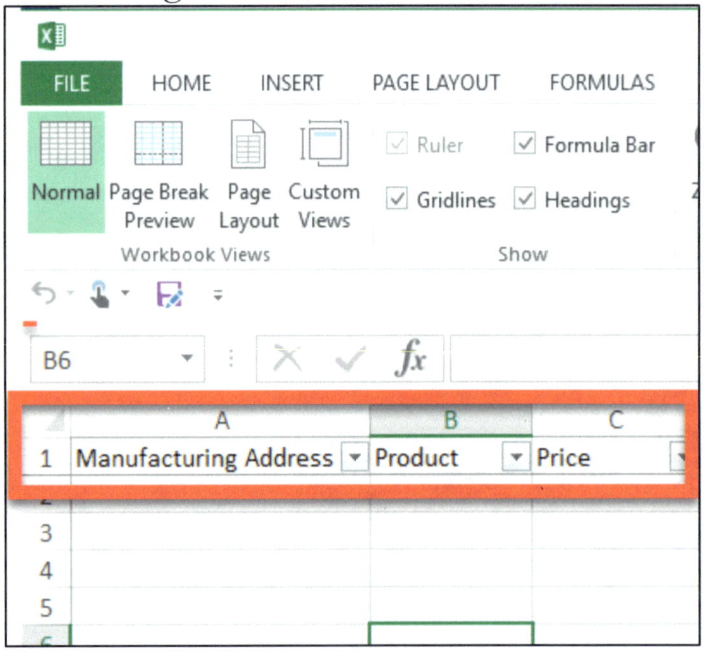

We will create the storage drop-down list of the companies in Cell K.

The formatting needs some help.

10. Select Column K.
11. On the Home tab, select **Wrap Text**.

When you wrap text, it displays the cell contents on multiple lines, rather than one long line.

The address still needs some formatting.

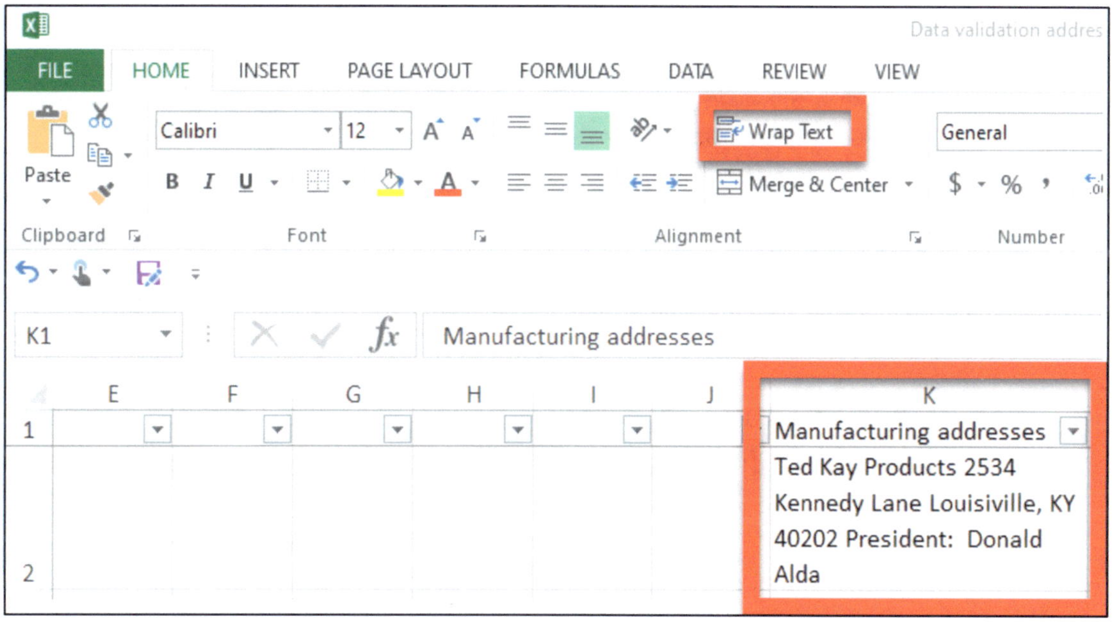

Note: This procedure changes the formatting of your storage column but not the drop-down list from which you select values. Once you create a drop-down list, you can't change its font size or styling. You may be able to do so with programming or another software program used in combination with Excel.

12. Now you want to click into Cell K2.
    Your address shows in the box above the column labels.

13. For this example, let's create line breaks within the cell by pressing **Alt-Enter** after the following text:

    Ted Kay Products

    2534 Kennedy Lane

    Louisville, KY 40202

    This address is formatted better; however, it could still be improved.

    Note that after you add a break in an address, it can be tricky to change anything afterwards. **Important Note**-To make changes after break: **Double Left Click, otherwise your cell will appear to be frozen.**

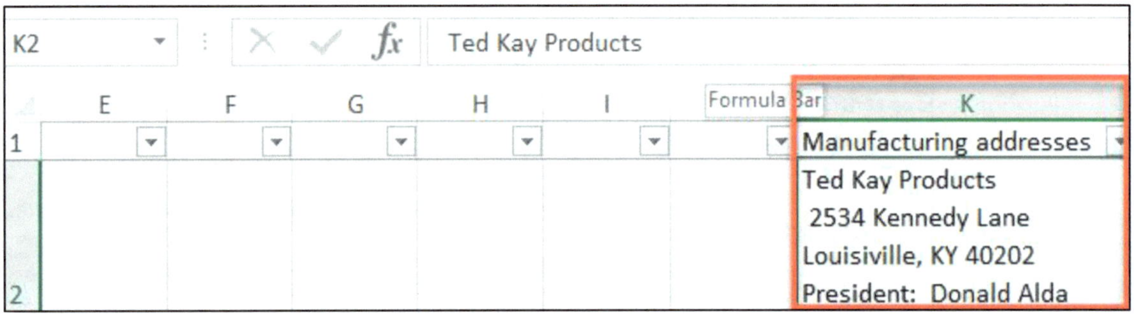

14. Select the words "President: Donald Alda."

15. Click B on the top menu bar to make the words "President Donald Alda" bold.

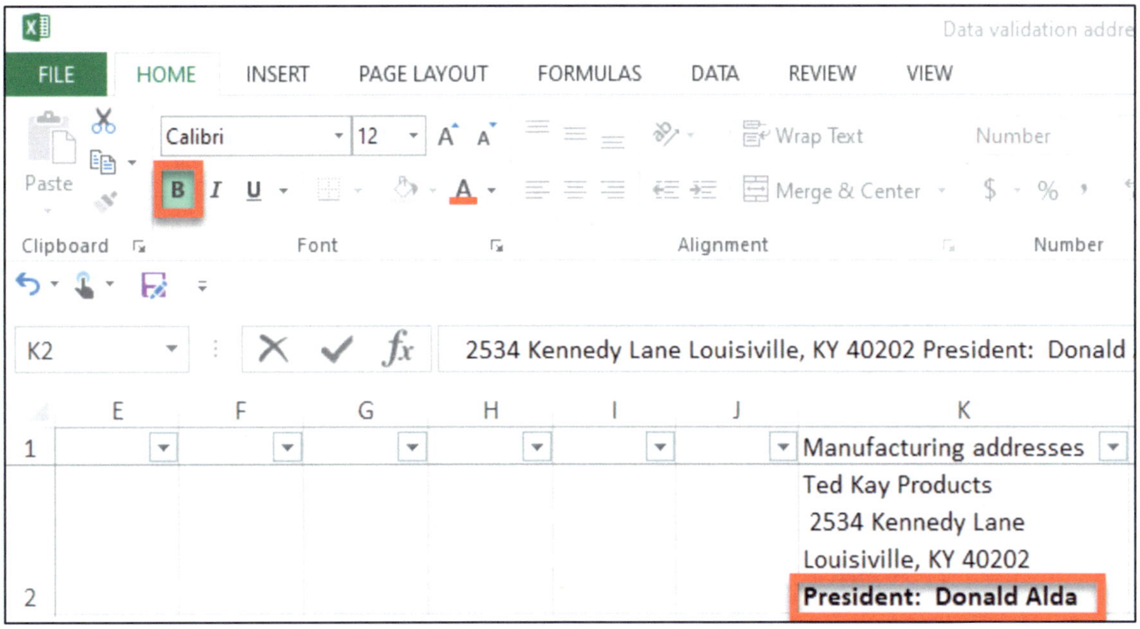

16. Select Ted Kay Products.
17. Change the font size to 14.
18. Click **B** to make the selected words bold.

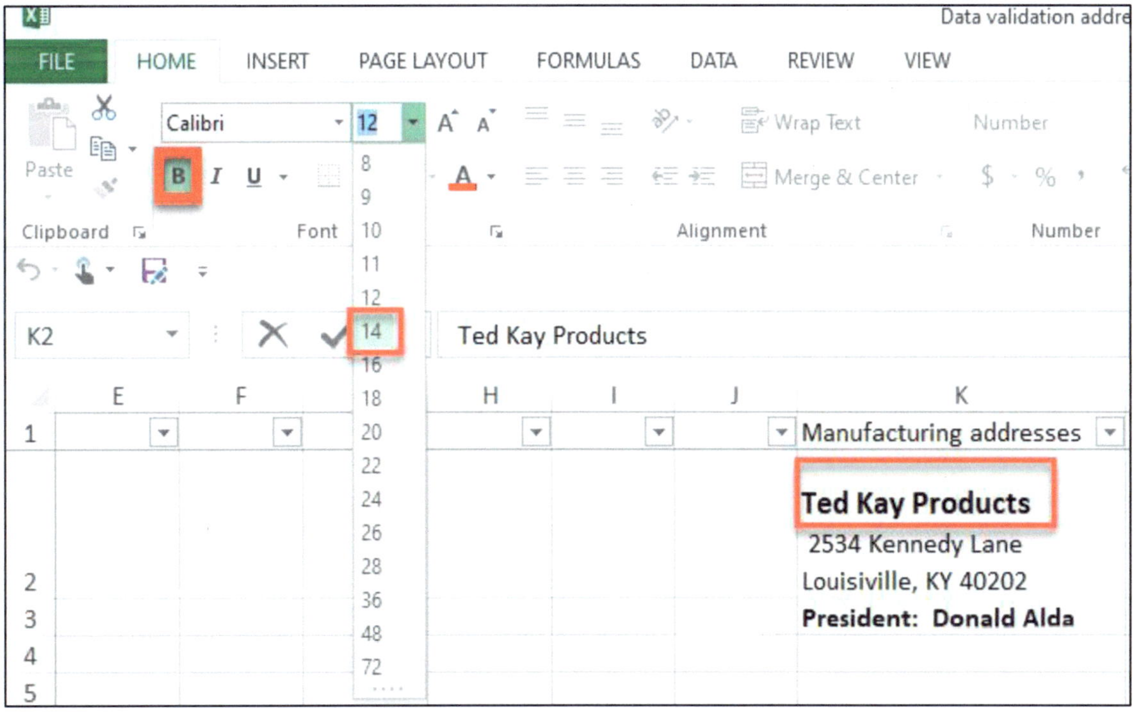

19. Press **Ctrl-S** to save the spreadsheet.

More Practice with Addresses

Let's add two more addresses to our spreadsheet.

1. In Cell K3, type in
   Bill Smith Manufacturing
   1300 Bayridge Drive
   Duluth, MN 55804
   President: Jessica Ebe
2. In Cell K4, type in
   Geof Bleeth Manufacturing
   698 Woodberry Rd.
   Stowe, VT 05672
   President: Sven Hanson

3. Refer to steps 13-18 of the "Creating the Spreadsheet with Addresses" procedure on page 76 to format these cells the same way you formatted "Ted Kay Products." It should look like this when you are done:

Note: This procedure changes the formatting of your storage column but not the drop-down list from which you select values. Once you create a drop-down list, you can't change its font size or styling. You may be able to do so with programming or another software program used in combination with Excel.

4. Select Column A because that is where the data validation will appear. Remember that the storage column is Column K.

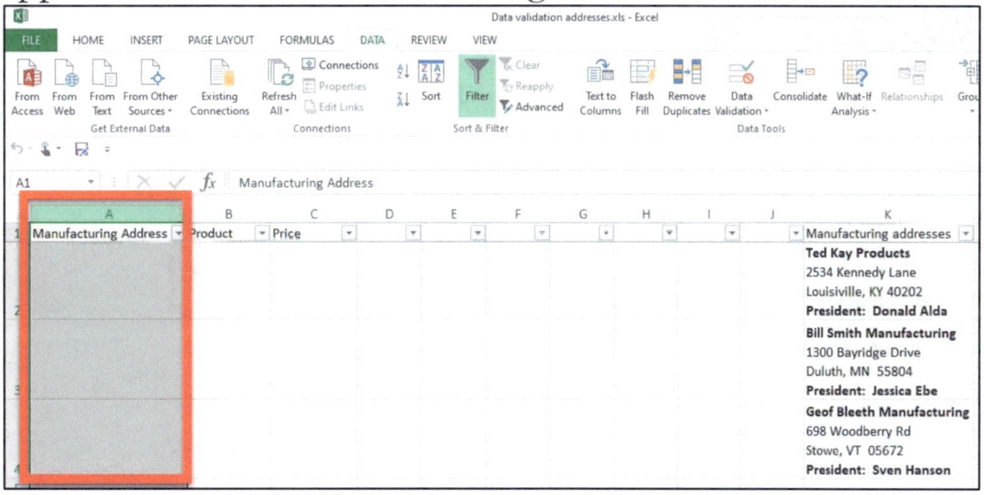

5. On the top menu bar, select **Data**, then 

, and then **Data Validation**. The Data Validation dialog box appears.

6. On the Settings tab in the Allow area, select **List**, and then click **OK**.

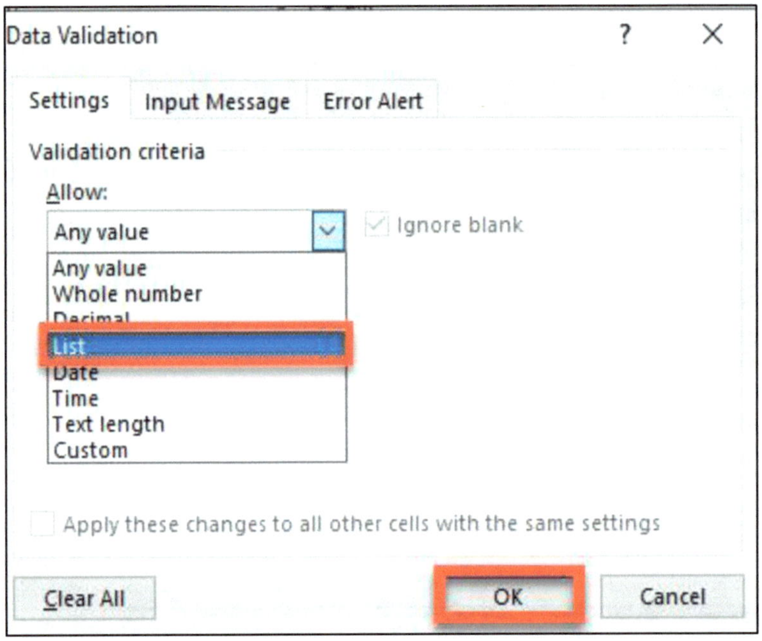

7. Click on the little square on the right side of the Source box.

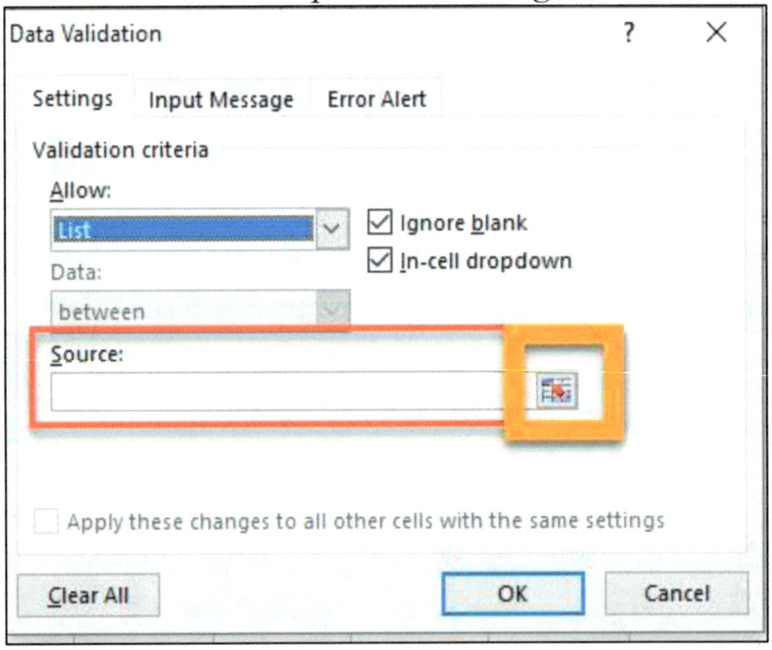

The Data Validation box appears.

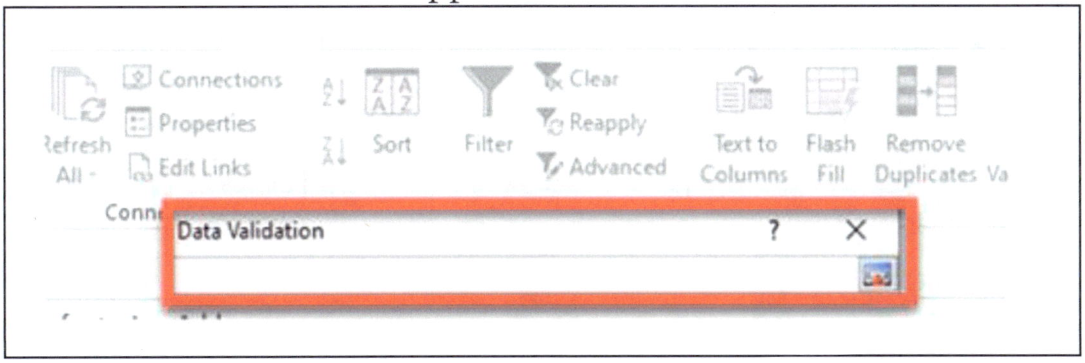

8. Click the first cell in the list in Cell K2, where you have the text you want to add to the drop-down.
9. Drag your cursor to the bottom of the list, to Cell K4, to select the whole list. You can see that a ***dashed line appears around the text*** you want in your drop-down.
10. Press **Enter**.

11. Click **OK**.
12. Click in Cell A2 and choose **Ted Kay Products** from the drop-down list.
13. Select Column A.

14. On the top menu bar, select the **Home** tab and then **Wrap Text**.

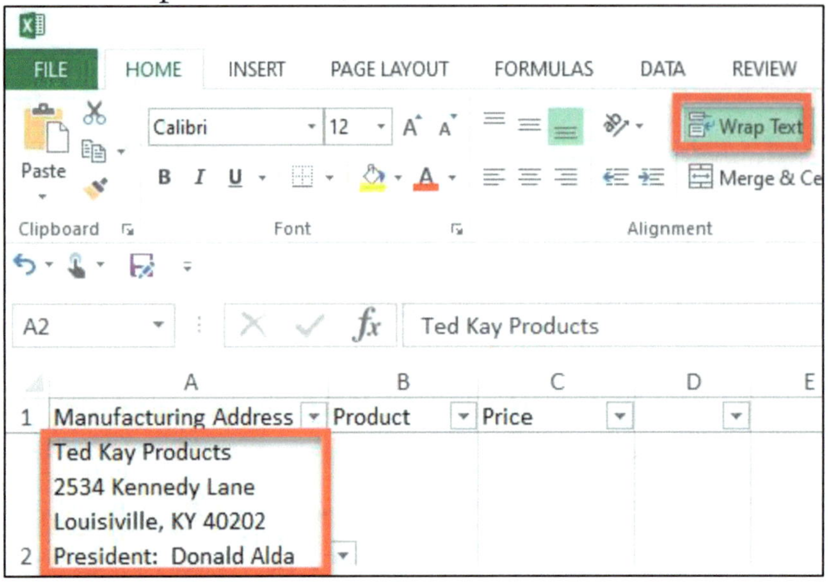

CHANGING A COMPANY NAME

When a company changes its name, data validation can be helpful.

You can input both company names in your spreadsheet.

Show the old company name with the new name.

Then you know both names of the company and avoid having duplicate companies in your list.

Companies
Healthy Vitamins
Paul Smith Vitamins
Paul Smith Vitamins ***Name changed to*** : Healthy Vitamins
Healthy Vitamins ***Formerly named*** : Paul Smith Vitamins

Break VS Wrap Text

BREAK VS WRAP TEXT			
	Appearance when Completed	**To Accomplish**	**Result**
Wrap Text	 	Top Ribbon Select: Home Tab and then select Wrap Text	Computer chooses how the text is formatted
Break	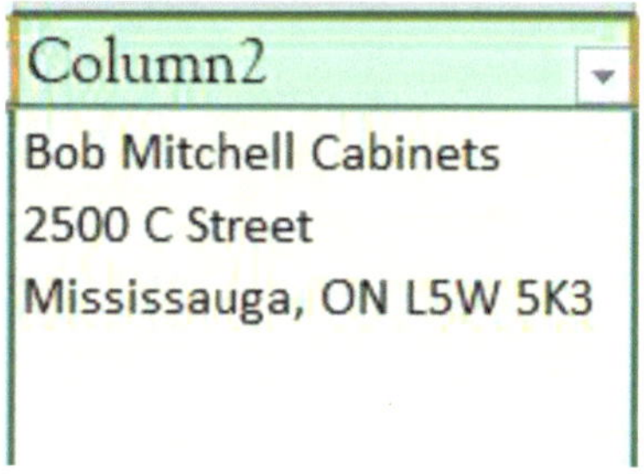 	Place your Cursor at the exact place where you want your break and then select **the ALT key** and then press **enter**.	You choose where the break will be in the text and how the text will be formatted.
Important Point About Break-after you a break you must left double click in order to change anything within the cell. If you do not do this,			

the cell will appear frozen.			

GLOSSARY

Term	Definition
Borders	The lines that define cells, rows, and columns in the spreadsheet. 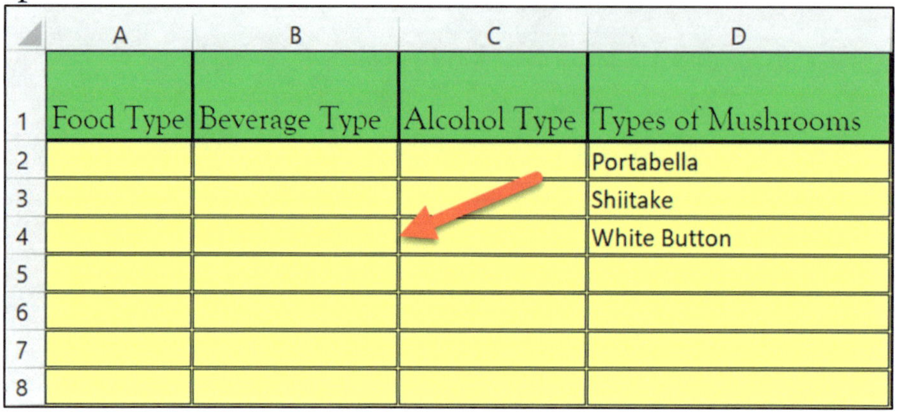
Data Validation Controls	These are controls that you use to create and manage your data validation drop-down list. 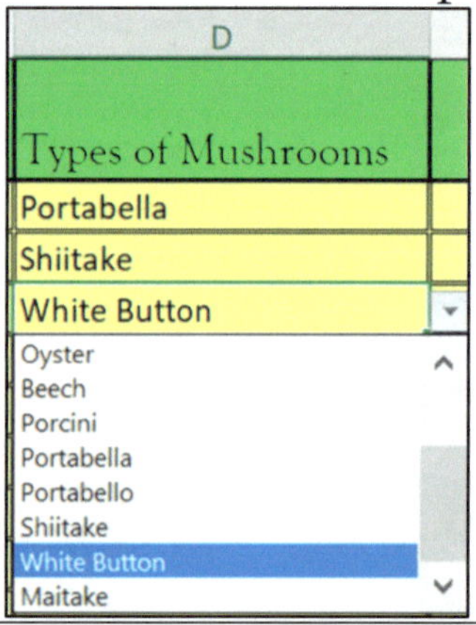
Header Row	Throughout this book, I refer to this row as the header row. 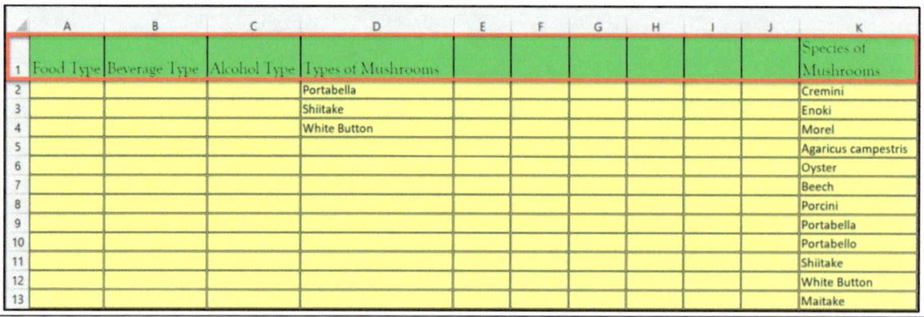

<table>
<tr><td>Storage Column</td><td>Throughout the book, this is Column K. This is the column where you enter the data validation information that will appear in your drop-down list.

</td></tr>
<tr><td>Top Menu Bar</td><td>The top menu bar gives you access to many of the operations in Excel. The menus are at the top and then each section below the menus include groups of icons that perform similar functions.
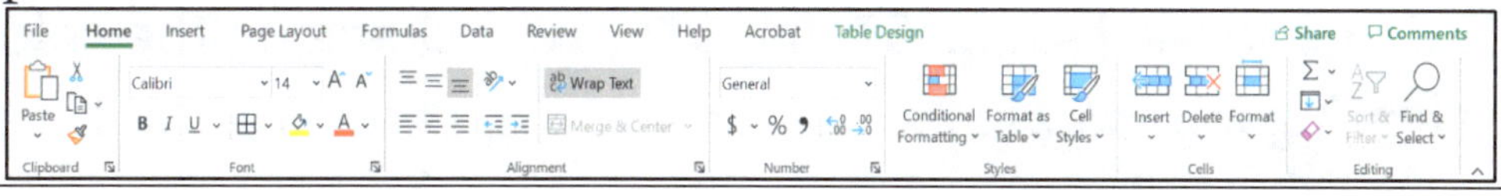
</td></tr>
</table>

Index

Made in the USA
Monee, IL
03 November 2026

67377025R10055